STATISTICAL INFERENCE
A Concise Course

Robert B. Ash
Department of Mathematics
University of Illinois

DOVER PUBLICATIONS, INC.
Mineola, New York

Bibliographical Note

This Dover edition, first published in 2011, is the first publication in book form of the work originally titled *Lectures on Statistics*, which is available from the author's website: http://www.math.uiuc.edu/~r-ash/

International Standard Book Number
ISBN-13: 978-0-486-48158-6
ISBN-10: 0-486-48158-1

Manufactured in the United States by Courier Corporation
48158101
www.doverpublications.com

Preface

This book is based on a course that I gave at UIUC in 1996 and again in 1997. No prior knowledge of statistics is assumed. A standard first course in probability is a prerequisite, but the first 8 lectures review results from basic probability that are important in statistics. Some exposure to matrix algebra is needed to cope with the multivariate normal distribution in Lecture 21, and there is a linear algebra review in Lecture 19. Here are the lecture titles:

Lecture 1. Transformation of Random Variables

Suppose we are given a random variable X with density $f_X(x)$. We apply a function g to produce a random variable $Y = g(X)$. We can think of X as the input to a black box, and Y the output. We wish to find the density or distribution function of Y. We illustrate the technique for the example in Figure 1.1.

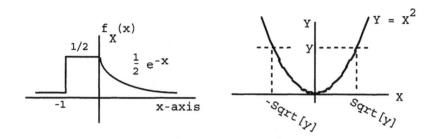

Figure 1.1

The **distribution function method** finds F_Y directly, and then f_Y by differentiation. We have $F_Y(y) = 0$ for $y < 0$. If $y \geq 0$, then $P\{Y \leq y\} = P\{-\sqrt{y} \leq x \leq \sqrt{y}\}$.

Case 1. $0 \leq y \leq 1$ (Figure 1.2). Then

$$F_Y(y) = \frac{1}{2}\sqrt{y} + \int_0^{\sqrt{y}} \frac{1}{2} e^{-x}\, dx = \frac{1}{2}\sqrt{y} + \frac{1}{2}(1 - e^{-\sqrt{y}}).$$

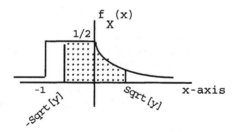

Figure 1.2

Case 2. $y > 1$ (Figure 1.3). Then

$$F_Y(y) = \frac{1}{2} + \int_0^{\sqrt{y}} \frac{1}{2} e^{-x}\, dx = \frac{1}{2} + \frac{1}{2}(1 - e^{-\sqrt{y}}).$$

The density of Y is 0 for $y < 0$ and

1

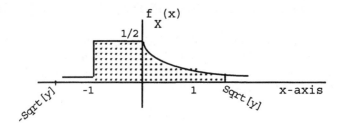

Figure 1.3

$$f_Y(y) = \frac{1}{4\sqrt{y}}(1 + e^{-\sqrt{y}}), \quad 0 < y < 1;$$

$$f_Y(y) = \frac{1}{4\sqrt{y}}e^{-\sqrt{y}}, \quad y > 1.$$

See Figure 1.4 for a sketch of f_Y and F_Y. (You can take $f_Y(y)$ to be anything you like at $y = 1$ because $\{Y = 1\}$ has probability zero.)

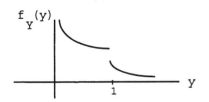
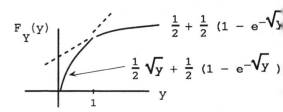

$$\frac{1}{2} + \frac{1}{2}(1 - e^{-\sqrt{y}})$$

$$\frac{1}{2}\sqrt{y} + \frac{1}{2}(1 - e^{-\sqrt{y}})$$

Figure 1.4

The **density function method** finds f_Y directly, and then F_Y by integration; see Figure 1.5. We have $f_Y(y)|dy| = f_X(\sqrt{y})dx + f_X(-\sqrt{y})dx$; we write $|dy|$ because probabilities are never negative. Thus

$$f_Y(y) = \frac{f_X(\sqrt{y})}{|dy/dx|_{x=\sqrt{y}}} + \frac{f_X(-\sqrt{y})}{|dy/dx|_{x=-\sqrt{y}}}$$

with $y = x^2$, $dy/dx = 2x$, so

$$f_Y(y) = \frac{f_X(\sqrt{y})}{2\sqrt{y}} + \frac{f_X(-\sqrt{y})}{2\sqrt{y}}.$$

(Note that $|-2\sqrt{y}| = 2\sqrt{y}$.) We have $f_Y(y) = 0$ for $y < 0$, and:
Case 1. $0 < y < 1$ (see Figure 1.2).

$$f_Y(y) = \frac{(1/2)e^{-\sqrt{y}}}{2\sqrt{y}} + \frac{1/2}{2\sqrt{y}} = \frac{1}{4}\sqrt{y}(1 + e^{-\sqrt{y}}).$$

Case 2. $y > 1$ (see Figure 1.3).

$$f_Y(y) = \frac{(1/2)e^{-\sqrt{y}}}{2\sqrt{y}} + 0 = \frac{1}{4\sqrt{y}}e^{-\sqrt{y}}$$

as before.

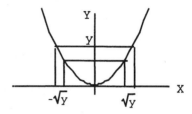

Figure 1.5

The distribution function method generalizes to situations where we have a single output but more than one input. For example, let X and Y be independent, each uniformly distributed on $[0, 1]$. The distribution function of $Z = X + Y$ is

$$F_Z(z) = P\{X + Y \leq z\} = \int\int_{x+y\leq z} f_{XY}(x, y)\, dx\, dy$$

with $f_{XY}(x, y) = f_X(x)f_Y(y)$ by independence. Now $F_Z(z) = 0$ for $z < 0$ and $F_Z(z) = 1$ for $z > 2$ (because $0 \leq Z \leq 2$).

Case 1. If $0 \leq z \leq 1$, then $F_Z(z)$ is the shaded area in Figure 1.6, which is $z^2/2$.

Case 2. If $1 \leq z \leq 2$, then $F_Z(z)$ is the shaded area in FIgure 1.7, which is $1 - [(2-z)^2/2]$. Thus (see Figure 1.8)

$$f_Z(z) = \begin{cases} z, & 0 \leq z \leq 1 \\ 2 - z, & 1 \leq z \leq 2 \, . \\ 0 & \text{elsewhere} \end{cases}$$

Problems

1. Let X, Y, Z be independent, identically distributed (from now on, abbreviated iid) random variables, each with density $f(x) = 6x^5$ for $0 \leq x \leq 1$, and 0 elsewhere. Find the distribution and density functions of the maximum of X, Y and Z.

2. Let X and Y be independent, each with density $e^{-x}, x \geq 0$. Find the distribution (from now on, an abbreviation for "Find the distribution or density function") of $Z = Y/X$.

3. A discrete random variable X takes values $x_1, \ldots, x_n$, each with probability $1/n$. Let $Y = g(X)$ where g is an arbitrary real-valued function. Express the probability function of Y ($p_Y(y) = P\{Y = y\}$) in terms of g and the x_i.

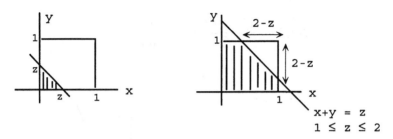

Figures 1.6 and 1.7

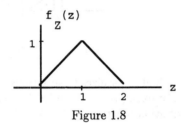

Figure 1.8

4. A random variable X has density $f(x) = ax^2$ on the interval $[0, b]$. Find the density of $Y = X^3$.

5. The *Cauchy density* is given by $f(y) = 1/[\pi(1 + y^2)]$ for all real y. Show that one way to produce this density is to take the tangent of a random variable X that is uniformly distributed between $-\pi/2$ and $\pi/2$.

Lecture 2. Jacobians

We need this idea to generalize the density function method to problems where there are k inputs and k outputs, with $k \geq 2$. However, if there are k inputs and $j < k$ outputs, often extra outputs can be introduced, as we will see later in the lecture.

2.1 The Setup

Let $X = X(U, V), Y = Y(U, V)$. Assume a one-to-one transformation, so that we can solve for U and V. Thus $U = U(X, Y), V = V(X, Y)$. Look at Figure 2.1. If u changes by du then x changes by $(\partial x/\partial u)\, du$ and y changes by $(\partial y/\partial u)\, du$. Similarly, if v changes by dv then x changes by $(\partial x/\partial v)\, dv$ and y changes by $(\partial y/\partial v)\, dv$. The small rectangle in the $u - v$ plane corresponds to a small parallelogram in the $x - y$ plane (Figure 2.2), with $A = (\partial x/\partial u, \partial y/\partial u, 0)\, du$ and $B = (\partial x/\partial v, \partial y/\partial v, 0)\, dv$. The area of the parallelogram is $|A \times B|$ and

$$A \times B = \begin{vmatrix} I & J & K \\ \partial x/\partial u & \partial y/\partial u & 0 \\ \partial x/\partial v & \partial y/\partial v & 0 \end{vmatrix} du\, dv = \begin{vmatrix} \partial x/\partial u & \partial x/\partial v \\ \partial y/\partial u & \partial y/\partial v \end{vmatrix} du\, dv\, K.$$

(A determinant is unchanged if we transpose the matrix, i.e., interchange rows and columns.)

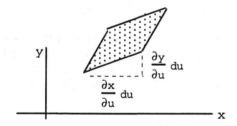 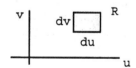

Figure 2.1

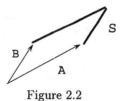

Figure 2.2

2.2 Definition and Discussion

The *Jacobian* of the transformation is

$$J = \begin{vmatrix} \partial x/\partial u & \partial x/\partial v \\ \partial y/\partial u & \partial y/\partial v \end{vmatrix}, \quad \text{written as} \quad \frac{\partial(x, y)}{\partial(u, v)}.$$

5

Thus $|A \times B| = |J| \, du \, dv$. Now $P\{(X,Y) \in S\} = P\{(U,V) \in R\}$, in other words, $f_{XY}(x,y)$ times the area of S is $f_{UV}(u,v)$ times the area of R. Thus

$$f_{XY}(x,y)|J| \, du \, dv = f_{UV}(u,v) \, du \, dv$$

and

$$f_{UV}(u,v) = f_{XY}(x,y)\left|\frac{\partial(x,y)}{\partial(u,v)}\right|.$$

The absolute value of the Jacobian $\partial(x,y)/\partial(u,v)$ gives a magnification factor for area in going from $u-v$ coordinates to $x-y$ coordinates. The magnification factor going the other way is $|\partial(u,v)/\partial(x,y)|$. But the magnification factor from $u-v$ to $u-v$ is 1, so

$$f_{UV}(u,v) = \left|\frac{f_{XY}(x,y)}{\partial(u,v)/\partial(x,y)}\right|.$$

In this formula, we must substitute $x = x(u,v), y = y(u,v)$ to express the final result in terms of u and v.

In three dimensions, a small rectangular box with volume $du \, dv \, dw$ corresponds to a parallelepiped in xyz space, determined by vectors

$$A = \left(\frac{\partial x}{\partial u} \quad \frac{\partial y}{\partial u} \quad \frac{\partial z}{\partial u}\right) du, \ B = \left(\frac{\partial x}{\partial v} \quad \frac{\partial y}{\partial v} \quad \frac{\partial z}{\partial v}\right) dv, \ C = \left(\frac{\partial x}{\partial w} \quad \frac{\partial y}{\partial w} \quad \frac{\partial z}{\partial w}\right) dw.$$

The volume of the parallelepiped is the absolute value of the dot product of A with $B \times C$, and the dot product can be written as a determinant with rows (or columns) A, B, C. This determinant is the Jacobian of x, y, z with respect to u, v, w [written $\partial(x,y,z)/\partial(u,v,w)$], times $du \, dv \, dw$. The volume magnification from uvw to xyz space is $|\partial(x,y,z)/\partial(u,v,w)|$ and we have

$$f_{UVW}(u,v,w) = \frac{f_{XYZ}(x,y,z)}{|\partial(u,v,w)/\partial(x,y,z)|}$$

with $x = x(u,v,w), y = y(u,v,w), z = z(u,v,w)$.

The jacobian technique extends to higher dimensions. The transformation formula is a natural generalization of the two and three-dimensional cases:

$$f_{Y_1 Y_2 \cdots Y_n}(y_1, \ldots, y_n) = \frac{f_{X_1 \cdots X_n}(x_1, \ldots, x_n)}{|\partial(y_1, \ldots, y_n)/\partial(x_1, \ldots, x_n)|}$$

where

$$\frac{\partial(y_1, \ldots, y_n)}{\partial(x_1, \ldots, x_n)} = \begin{vmatrix} \frac{\partial y_1}{\partial x_1} & \cdots & \frac{\partial y_1}{\partial x_n} \\ & \vdots & \\ \frac{\partial y_n}{\partial x_1} & \cdots & \frac{\partial y_n}{\partial x_n} \end{vmatrix}.$$

To help you remember the formula, think $f(y) \, dy = f(x) dx$.

2.3 A Typical Application

Let X and Y be independent, positive random variables with densities f_X and f_Y, and let $Z = XY$. We find the density of Z by introducing a new random variable W, as follows:

$$Z = XY, \quad W = Y$$

($W = X$ would be equally good). The transformation is one-to-one because we can solve for X, Y in terms of Z, W by $X = Z/W, Y = W$. In a problem of this type, we must always pay attention to the range of the variables: $x > 0, y > 0$ is equivalent to $z > 0, w > 0$. Now

$$f_{ZW}(z, w) = \frac{f_{XY}(x, y)}{|\partial(z, w)/\partial(x, y)|_{x=z/w, y=w}}$$

with

$$\frac{\partial(z, w)}{\partial(x, y)} = \begin{vmatrix} \partial z/\partial x & \partial z/\partial y \\ \partial w/\partial x & \partial w/\partial y \end{vmatrix} = \begin{vmatrix} y & x \\ 0 & 1 \end{vmatrix} = y.$$

Thus

$$f_{ZW}(z, w) = \frac{f_X(x) f_Y(y)}{w} = \frac{f_X(z/w) f_Y(w)}{w}$$

and we are left with the problem of finding the marginal density from a joint density:

$$f_Z(z) = \int_{-\infty}^{\infty} f_{ZW}(z, w) \, dw = \int_0^{\infty} \frac{1}{w} f_X(z/w) f_Y(w) \, dw.$$

Problems

1. The joint density of two random variables X_1 and X_2 is $f(x_1, x_2) = 2e^{-x_1} e^{-x_2}$, where $0 < x_1 < x_2 < \infty$; $f(x_1, x_2) = 0$ elsewhere. Consider the transformation $Y_1 = 2X_1$, $Y_2 = X_2 - X_1$. Find the joint density of Y_1 and Y_2, and conclude that Y_1 and Y_2 are independent.

2. Repeat Problem 1 with the following new data. The joint density is given by $f(x_1, x_2) = 8x_1 x_2$, $0 < x_1 < x_2 < 1$; $f(x_1, x_2) = 0$ elsewhere; $Y_1 = X_1/X_2$, $Y_2 = X_2$.

3. Repeat Problem 1 with the following new data. We now have three iid random variables $X_i, i = 1, 2, 3$, each with density $e^{-x}, x > 0$. The transformation equations are given by $Y_1 = X_1/(X_1 + X_2)$, $Y_2 = (X_1 + X_2)/(X_1 + X_2 + X_3)$, $Y_3 = X_1 + X_2 + X_3$. As before, find the joint density of the Y_i and show that Y_1, Y_2 and Y_3 are independent.

Comments on the Problem Set

In Problem 3, notice that $Y_1 Y_2 Y_3 = X_1$, $Y_2 Y_3 = X_1 + X_2$, so $X_2 = Y_2 Y_3 - Y_1 Y_2 Y_3$, $X_3 = (X_1 + X_2 + X_3) - (X_1 + X_2) = Y_3 - Y_2 Y_3$.

If $f_{XY}(x, y) = g(x)h(y)$ for all x, y, then X and Y are independent, because

$$f(y|x) = \frac{f_{XY}(x, y)}{f_X(x)} = \frac{g(x)h(y)}{g(x) \int_{-\infty}^{\infty} h(y) \, dy}$$

which does not depend on x. The set of points where $g(x) = 0$ (equivalently $f_X(x) = 0$) can be ignored because it has probability zero. It is important to realize that in this argument, "for all x, y" means that x and y must be allowed to vary independently of each other, so the set of possible x and y must be of the rectangular form $a < x < b, c < y < d$. (The constants a, b, c, d can be infinite.) For example, if $f_{XY}(x, y) = 2e^{-x}e^{-y}, 0 < y < x$, and 0 elsewhere, then X and Y are *not* independent. Knowing x forces $0 < y < x$, so the conditional distribution of Y given $X = x$ certainly depends on x. Note that $f_{XY}(x, y)$ is *not* a function of x alone times a function of y alone. We have

$$f_{XY}(x, y) = 2e^{-x}e^{-y}I[0 < y < x]$$

where the *indicator* I is 1 for $0 < y < x$ and 0 elsewhere.

In Jacobian problems, pay close attention to the range of the variables. For example, in Problem 1 we have $y_1 = 2x_1, y_2 = x_2 - x_1$, so $x_1 = y_1/2, x_2 = (y_1/2) + y_2$. From these equations it follows that $0 < x_1 < x_2 < \infty$ is equivalent to $y_1 > 0, y_2 > 0$.

Lecture 3. Moment-Generating Functions

3.1 Definition

The *moment-generating function* of a random variable X is defined by

$$M(t) = M_X(t) = E[e^{tX}]$$

where t is a real number. To see the reason for the terminology, note that $M(t)$ is the expectation of $1 + tX + t^2 X^2/2! + t^3 X^3/3! + \cdots$. If $\mu_n = E(X^n)$, the n-th moment of X, and we can take the expectation term by term, then

$$M(t) = 1 + \mu_1 t + \frac{\mu_2 t^2}{2!} + \cdots + \frac{\mu_n t^n}{n!} + \cdots.$$

Since the coefficient of t^n in the Taylor expansion is $M^{(n)}(0)/n!$, where $M^{(n)}$ is the n-th derivative of M, we have $\mu_n = M^{(n)}(0)$.

3.2 The Key Theorem

If $Y = \sum_{i=1}^{n} X_i$ where $X_1, \ldots, X_n$ are independent, then $M_Y(t) = \prod_{i=1}^{n} M_{X_i}(t)$.

Proof. First note that if X and Y are independent, then

$$E[g(X)h(Y)] = \int_{-\infty}^{\infty} \int_{-\infty}^{\infty} g(x)h(y) f_{XY}(x, y) \, dx \, dy.$$

Since $f_{XY}(x, y) = f_X(x) f_Y(y)$, the double integral becomes

$$\int_{-\infty}^{\infty} g(x) f_X(x) \, dx \int_{-\infty}^{\infty} h(y) f_Y(y) \, dy = E[g(X)]E[h(Y)]$$

and similarly for more than two random variables. Now if $Y = X_1 + \cdots + X_n$ with the X_i's independent, we have

$$M_Y(t) = E[e^{tY}] = E[e^{tX_1} \cdots e^{tX_n}] = E[e^{tX_1}] \cdots E[e^{tX_n}] = M_{X_1}(t) \cdots M_{X_n}(t). \quad \clubsuit$$

3.3 The Main Application

Given independent random variables $X_1, \ldots, X_n$ with densities $f_1, \ldots, f_n$ respectively, find the density of $Y = \sum_{i=1}^{n} X_i$.

Step 1. Compute $M_i(t)$, the moment-generating function of X_i, for each i.

Step 2. Compute $M_Y(t) = \prod_{i=1}^{n} M_i(t)$.

Step 3. From $M_Y(t)$ find $f_Y(y)$.

This technique is known as a *transform method*. Notice that the moment-generating function and the density of a random variable are related by $M(t) = \int_{-\infty}^{\infty} e^{tx} f(x) \, dx$. With t replaced by $-s$ we have a *Laplace transform*, and with t replaced by it we have a *Fourier transform*. The strategy works because at step 3, the moment-generating function determines the density uniquely. (This is a theorem from Laplace or Fourier transform theory.)

9

3.4 Examples

1. *Bernoulli Trials.* Let X be the number of successes in n trials with probability of success p on a given trial. Then $X = X_1 + \cdots + X_n$, where $X_i = 1$ if there is a success on trial i and $X_i = 0$ if there is a failure on trial i. Thus

$$M_i(t) = E[e^{tX_i}] = P\{X_i = 1\}e^{t1} + P\{X_i = 0\}e^{t0} = pe^t + q$$

with $p + q = 1$. The moment-generating function of X is

$$M_X(t) = (pe^t + q)^n = \sum_{k=0}^{n} \binom{n}{k} p^k q^{n-k} e^{tk}.$$

This could have been derived directly:

$$M_X(t) = E[e^{tX}] = \sum_{k=0}^{n} P\{X = k\}e^{tk} = \sum_{k=0}^{n} \binom{n}{k} p^k q^{n-k} e^{tk} = (pe^t + q)^n$$

by the binomial theorem.

2. *Poisson.* We have $P\{X = k\} = e^{-\lambda}\lambda^k/k!, \quad k = 0, 1, 2, \ldots$. Thus

$$M(t) = \sum_{k=0}^{\infty} \frac{e^{-\lambda}\lambda^k}{k!} e^{tk} = e^{-\lambda} \sum_{k=0}^{\infty} \frac{(\lambda e^t)^k}{k!} = \exp(-\lambda)\exp(\lambda e^t) = \exp[\lambda(e^t - 1)].$$

We can compute the mean and variance from the moment-generating function:

$$E(X) = M'(0) = [\exp(\lambda(e^t - 1))\lambda e^t]_{t=0} = \lambda.$$

Let $h(\lambda, t) = \exp[\lambda(e^t - 1)]$. Then

$$E(X^2) = M''(0) = [h(\lambda, t)\lambda e^t + \lambda e^t h(\lambda, t)\lambda e^t]_{t=0} = \lambda + \lambda^2$$

hence

$$\text{Var}\, X = E(X^2) - [E(X)]^2 = \lambda + \lambda^2 - \lambda^2 = \lambda.$$

3. *Normal(0,1).* The moment-generating function is

$$M(t) = E[e^{tX}] = \int_{-\infty}^{\infty} e^{tx} \frac{1}{\sqrt{2\pi}} e^{-x^2/2}\, dx$$

Now $-(x^2/2) + tx = -(1/2)(x^2 - 2tx + t^2 - t^2) = -(1/2)(x - t)^2 + (1/2)t^2$ so

$$M(t) = e^{t^2/2} \int_{-\infty}^{\infty} \frac{1}{\sqrt{2\pi}} \exp[-(x - t)^2/2]\, dx.$$

The integral is the area under a normal density (mean t, variance 1), which is 1. Consequently,

$$M(t) = e^{t^2/2}.$$

4. *Normal*(μ, σ^2). If X is normal(μ, σ^2), then $Y = (X - \mu)/\sigma$ is normal(0,1). This is a good application of the density function method from Lecture 1:

$$f_Y(y) = \frac{f_X(x)}{|dy/dx|_{x=\mu+\sigma y}} = \sigma \frac{1}{\sqrt{2\pi}\sigma} e^{-y^2/2}.$$

We have $X = \mu + \sigma Y$, so

$$M_X(t) = E[e^{tX}] = e^{t\mu} E[e^{t\sigma Y}] = e^{t\mu} M_Y(t\sigma).$$

Thus

$$M_X(t) = e^{t\mu} e^{t^2\sigma^2/2}.$$

Remember this technique, which is especially useful when $Y = aX + b$ and the moment-generating function of X is known.

3.5 Theorem

If X is normal(μ, σ^2) and $Y = aX + b$, then Y is normal$(a\mu + b, a^2\sigma^2)$.
Proof. We compute

$$M_Y(t) = E[e^{tY}] = E[e^{t(aX+b)}] = e^{bt} M_X(at) = e^{bt} e^{at\mu} e^{a^2 t^2 \sigma^2/2}.$$

Thus

$$M_Y(t) = \exp[t(a\mu + b)] \exp(t^2 a^2 \sigma^2/2). \quad \clubsuit$$

Here is another basic result.

3.6 Theorem

Let $X_1, \ldots, X_n$ be independent, with X_i normal (μ_i, σ_i^2). Then $Y = \sum_{i=1}^n X_i$ is normal with mean $\mu = \sum_{i=1}^n \mu_i$ and variance $\sigma^2 = \sum_{i=1}^n \sigma_i^2$.
Proof. The moment-generating function of Y is

$$M_Y(t) = \prod_{i=1}^n \exp(t_i \mu_i + t^2 \sigma_i^2/2) = \exp(t\mu + t^2\sigma^2/2). \quad \clubsuit$$

A similar argument works for the Poisson distribution; see Problem 4.

3.7 The Gamma Distribution

First, we define the *gamma function* $\Gamma(\alpha) = \int_0^\infty y^{\alpha-1} e^{-y}\,dy$, $\alpha > 0$. We need three properties:

(a) $\Gamma(\alpha + 1) = \alpha\Gamma(\alpha)$, the *recursion formula*;
(b) $\Gamma(n + 1) = n!, n = 0, 1, 2, \ldots$;

(c) $\Gamma(1/2) = \sqrt{\pi}$.

To prove (a), integrate by parts: $\Gamma(\alpha) = \int_0^\infty e^{-y}d(y^\alpha/\alpha)$. Part (b) is a special case of (a). For (c) we make the change of variable $y = z^2/2$ and compute

$$\Gamma(1/2) = \int_0^\infty y^{-1/2}e^{-y}\,dy = \int_0^\infty \sqrt{2}z^{-1}e^{-z^2/2}z\,dz.$$

The second integral is $2\sqrt{\pi}$ times half the area under the normal(0,1) density, that is, $2\sqrt{\pi}(1/2) = \sqrt{\pi}$.

The *gamma density* is

$$f(x) = \frac{1}{\Gamma(\alpha)\beta^\alpha}x^{\alpha-1}e^{-x/\beta}$$

where α and β are positive constants. The moment-generating function is

$$M(t) = \int_0^\infty [\Gamma(\alpha)\beta^\alpha]^{-1}x^{\alpha-1}e^{tx}e^{-x/\beta}\,dx.$$

Change variables via $y = (-t + (1/\beta))x$ to get

$$\int_0^\infty [\Gamma(\alpha)\beta^\alpha]^{-1}\left(\frac{y}{-t+(1/\beta)}\right)^{\alpha-1}e^{-y}\frac{dy}{-t+(1/\beta)}$$

which reduces to

$$\frac{1}{\beta^\alpha}\left(\frac{\beta}{1-\beta t}\right)^\alpha = (1-\beta t)^{-\alpha}.$$

In this argument, t must be less than $1/\beta$ so that the integrals will be finite.

Since $M(0) = \int_{-\infty}^\infty f(x)\,dx = \int_0^\infty f(x)\,dx$ in this case, with $f \geq 0$, $M(0) = 1$ implies that we have a legal probability density. As before, moments can be calculated efficiently from the moment-generating function:

$$E(X) = M'(0) = -\alpha(1-\beta t)^{-\alpha-1}(-\beta)|_{t=0} = \alpha\beta;$$

$$E(X^2) = M''(0) = -\alpha(-\alpha-1)(1-\beta t)^{-\alpha-2}(-\beta)^2|_{t=0} = \alpha(\alpha+1)\beta^2.$$

Thus

$$\text{Var } X = E(X^2) - [E(X)]^2 = \alpha\beta^2.$$

3.8 Special Cases

The *exponential density* is a gamma density with $\alpha = 1$: $f(x) = (1/\beta)e^{-x/\beta}, x \geq 0$, with $E(X) = \beta$, $E(X^2) = 2\beta^2$, $\text{Var } X = \beta^2$.

A random variable X has the *chi square density* with r *degrees of freedom* ($X = \chi^2(r)$ for short, where r is a positive integer) if its density is gamma with $\alpha = r/2$ and $\beta = 2$. Thus

$$f(x) = \frac{1}{\Gamma(r/2)2^{r/2}} x^{(r/2)-1} e^{-x/2}, \quad x \geq 0$$

and

$$M(t) = \frac{1}{(1 - 2t)^{r/2}}, \quad t < 1/2.$$

Therefore $E[\chi^2(r)] = \alpha\beta = r$, $\quad \text{Var}[\chi^2(r)] = \alpha\beta^2 = 2r$.

3.9 Lemma

If X is normal(0,1) then X^2 is $\chi^2(1)$.

Proof. We compute the moment-generating function of X^2 directly:

$$M_{X^2}(t) = E[e^{tX^2}] = \int_{-\infty}^{\infty} e^{tx^2} \frac{1}{\sqrt{2\pi}} e^{-x^2/2} \, dx.$$

Let $y = \sqrt{1 - 2t}x$; the integral becomes

$$\int_{-\infty}^{\infty} \frac{1}{\sqrt{2\pi}} e^{-y^2/2} \frac{dy}{\sqrt{1 - 2t}} = (1 - 2t)^{-1/2}$$

which is $\chi^2(1)$. ♣

3.10 Theorem

If $X_1, \ldots, X_n$ are independent, each normal $(0,1)$, then $Y = \sum_{i=1}^{n} X_i^2$ is $\chi^2(n)$.

Proof. By (3.9), each X_i^2 is $\chi^2(1)$ with moment-generating function $(1 - 2t)^{-1/2}$. Thus $M_Y(t) = (1 - 2t)^{-n/2}$ for $t < 1/2$, which is $\chi^2(n)$. ♣

3.11 Another Method

Another way to find the density of $Z = X + Y$ where X and Y are independent random variables is by the *convolution formula*

$$f_Z(z) = \int_{-\infty}^{\infty} f_X(x) f_Y(z - x) \, dx = \int_{-\infty}^{\infty} f_Y(y) f_X(z - y) \, dy.$$

To see this intuitively, reason as follows. The probability that Z lies near z (between z and $z + dz$) is $f_Z(z) \, dz$. Let us compute this in terms of X and Y. The probability that X lies near x is $f_X(x) \, dx$. Given that X lies near x, Z will lie near z if and only if Y lies near $z - x$, in other words, $z - x \leq Y \leq z - x + dz$. By independence of X and Y, this probability is $f_Y(z-x) \, dz$. Thus $f_Z(z)$ is a sum of terms of the form $f_X(x) \, dx \, f_Y(z-x) \, dz$. Cancel the dz's and replace the sum by an integral to get the result. A formal proof can be given using Jacobians.

3.12 The Poisson Process

This process occurs in many physical situations, and provides an application of the gamma distribution. For example, particles can arrive at a counting device, customers at a serving counter, airplanes at an airport, or phone calls at a telephone exchange. Divide the time interval $[0, t]$ into a large number n of small subintervals of length dt, so that $n\,dt = t$. If $I_i, i = 1, \dots, n$, is one of the small subintervals, we make the following assumptions:

(1) The probability of exactly one arrival in I_i is $\lambda\,dt$, where λ is a constant.

(2) The probability of no arrivals in I_i is $1 - \lambda\,dt$.

(3) The probability of more than one arrival in I_i is zero.

(4) If A_i is the event of an arrival in I_i, then the $A_i, i = 1, \dots, n$ are independent.

As a consequence of these assumptions, we have $n = t/dt$ Bernoulli trials with probability of success $p = \lambda\,dt$ on a given trial. As $dt \to 0$ we have $n \to \infty$ and $p \to 0$, with $np = \lambda t$. We conclude that the number $N[0, t]$ of arrivals in $[0, t]$ is Poisson (λt):

$$P\{N[0, t] = k\} = e^{-\lambda t}(\lambda t)^k/k!, k = 0, 1, 2, \dots .$$

Since $E(N[0, t]) = \lambda t$, we may interpret λ as the *average number of arrivals per unit time*.

Now let W_1 be the waiting time for the first arrival. Then

$$P\{W_1 > t\} = P\{\text{no arrival in } [0, t]\} = P\{N[0, t] = 0\} = e^{-\lambda t}, t \geq 0.$$

Thus $F_{W_1}(t) = 1 - e^{-\lambda t}$ and $f_{W_1}(t) = \lambda e^{-\lambda t}, t \geq 0$. From the formulas for the mean and variance of an exponential random variable we have $E(W_1) = 1/\lambda$ and $\text{Var}\, W_1 = 1/\lambda^2$.

Let W_k be the (total) waiting time for the k-th arrival. Then W_k is the waiting time for the first arrival plus the time after the first up to the second arrival plus $\cdots$ plus the time after arrival $k - 1$ up to the k-th arrival. Thus W_k is the sum of k independent exponential random variables, and

$$M_{W_k}(t) = \left(\frac{1}{1 - (t/\lambda)}\right)^k$$

so W_k is gamma with $\alpha = k, \beta = 1/\lambda$. Therefore

$$f_{W_k}(t) = \frac{1}{(k-1)!}\lambda^k t^{k-1} e^{-\lambda t}, t \geq 0.$$

Problems

1. Let X_1 and X_2 be independent, and assume that X_1 is $\chi^2(r_1)$ and $Y = X_1 + X_2$ is $\chi^2(r)$, where $r > r_1$. Show that X_2 is $\chi^2(r_2)$, where $r_2 = r - r_1$.

2. Let X_1 and X_2 be independent, with X_i gamma with parameters α_i and $\beta_i, i = 1, 2$. If c_1 and c_2 are positive constants, find convenient sufficient conditions under which $c_1 X_1 + c_2 X_2$ will also have the gamma distribution.

3. If $X_1, \dots, X_n$ are independent random variables with moment-generating functions $M_1, \dots, M_n$, and $c_1, \dots, c_n$ are constants, express the moment-generating function M of $c_1 X_1 + \cdots + c_n X_n$ in terms of the M_i.

4. If $X_1, \ldots, X_n$ are independent, with X_i Poisson$(\lambda_i), i = 1, \ldots, n$, show that the sum $Y = \sum_{i=1}^n X_i$ has the Poisson distribution with parameter $\lambda = \sum_{i=1}^n \lambda_i$.

5. An unbiased coin is tossed independently n_1 times and then again tossed independently n_2 times. Let X_1 be the number of heads in the first experiment, and X_2 the number of *tails* in the second experiment. Without using moment-generating functions, in fact without any calculation at all, find the distribution of $X_1 + X_2$.

Lecture 4. Sampling From a Normal Population

4.1 Definitions and Comments

Let $X_1, \dots, X_n$ be iid. The *sample mean* of the X_i is

$$\overline{X} = \frac{1}{n} \sum_{i=1}^{n} X_i$$

and the *sample variance* is

$$S^2 = \frac{1}{n} \sum_{i=1}^{n} (X_i - \overline{X})^2.$$

If the X_i have mean μ and variance σ^2, then

$$E(\overline{X}) = \frac{1}{n} \sum_{i=1}^{n} E(X_i) = \frac{1}{n} n\mu = \mu$$

and

$$\operatorname{Var} \overline{X} = \frac{1}{n^2} \sum_{i=1}^{n} \operatorname{Var} X_i = \frac{n\sigma^2}{n^2} = \frac{\sigma^2}{n} \to 0 \quad \text{as} \quad n \to \infty.$$

Thus $\overline{X}$ is a good estimate of μ. (For large n, the variance of $\overline{X}$ is small, so $\overline{X}$ is concentrated near its mean.) The sample variance is an average squared deviation from the sample mean, but it is a biased estimate of the true variance σ^2:

$$E[(X_i - \overline{X})^2] = E[(X_i - \mu) - (\overline{X} - \mu)]^2 = \operatorname{Var} X_i + \operatorname{Var} \overline{X} - 2E[(X_i - \mu)(\overline{X} - \mu)].$$

Notice the *centralizing technique*. We subtract and add back the mean of X_i, which will make the cross terms easier to handle when squaring. The above expression simplifies to

$$\sigma^2 + \frac{\sigma^2}{n} - 2E\left[(X_i - \mu)\frac{1}{n} \sum_{j=1}^{n} (X_j - \mu)\right] = \sigma^2 + \frac{\sigma^2}{n} - \frac{2}{n} E[(X_i - \mu)^2].$$

Thus

$$E[(X_i - \overline{X})^2] = \sigma^2 \left(1 + \frac{1}{n} - \frac{2}{n}\right) = \frac{n-1}{n} \sigma^2.$$

Consequently, $E(S^2) = (n-1)\sigma^2/n$, not σ^2. Some books define the sample variance as

$$\frac{1}{n-1} \sum_{i=1}^{n} (X_i - \overline{X})^2 = \frac{n}{n-1} S^2$$

where S^2 is our sample variance. This adjusted estimate of the true variance is unbiased (its expectation is σ^2), but *biased does not mean bad*. If we measure performance by asking for a small mean square error, the biased estimate is better in the normal case, as we will see at the end of the lecture.

17

4.2 The Normal Case

We now assume that the X_i are normally distributed, and find the distribution of S^2. Let $y_1 = \bar{x} = (x_1 + \cdots + x_n)/n$, $\quad y_2 = x_2 - \bar{x}, \ldots, y_n = x_n - \bar{x}$. Then $y_1 + y_2 = x_2$, $\quad y_1 + y_3 = x_3, \ldots, y_1 + y_n = x_n$. Add these equations to get $(n-1)y_1 + y_2 + \cdots + y_n = x_2 + \cdots + x_n$, or

$$ny_1 + (y_2 + \cdots + y_n) = (x_2 + \cdots + x_n) + y_1 \tag{1}$$

But $ny_1 = n\bar{x} = x_1 + \cdots + x_n$, so by cancelling $x_2, \ldots, x_n$ in (1), $x_1 + (y_2 + \cdots + y_n) = y_1$. Thus we can solve for the x's in terms of the y's:

$$\begin{aligned} x_1 &= y_1 - y_2 - \cdots - y_n \\ x_2 &= y_1 + y_2 \\ x_3 &= y_1 + y_3 \\ &\vdots \\ x_n &= y_1 + y_n \end{aligned} \tag{2}$$

The Jacobian of the transformation is

$$d_n = \frac{\partial(x_1, \ldots, x_n)}{\partial(y_1, \ldots, y_n)} = \begin{vmatrix} 1 & -1 & -1 & \cdots & -1 \\ 1 & 1 & 0 & \cdots & 0 \\ 1 & 0 & 1 & \cdots & 0 \\ \vdots & & & & \\ 1 & 0 & 0 & \cdots & 1 \end{vmatrix}$$

To see the pattern, look at the 4 by 4 case and expand via the last row:

$$\begin{vmatrix} 1 & -1 & -1 & -1 \\ 1 & 1 & 0 & 0 \\ 1 & 0 & 1 & 0 \\ 1 & 0 & 0 & 1 \end{vmatrix} = (-1) \begin{vmatrix} -1 & -1 & -1 \\ 1 & 0 & 0 \\ 0 & 1 & 0 \end{vmatrix} + \begin{vmatrix} 1 & -1 & -1 \\ 1 & 1 & 0 \\ 1 & 0 & 1 \end{vmatrix}$$

so $d_4 = 1 + d_3$. In general, $d_n = 1 + d_{n-1}$, and since $d_2 = 2$ by inspection, we have $d_n = n$ for all $n \geq 2$. Now

$$\sum_{i=1}^{n}(x_i - \mu)^2 = \sum(x_i - \bar{x} + \bar{x} - \mu)^2 = \sum(x_i - \bar{x})^2 + n(\bar{x} - \mu)^2 \tag{3}$$

because $\sum(x_i - \bar{x}) = 0$. By (2), $x_1 - \bar{x} = x_1 - y_1 = -y_2 - \cdots - y_n$ and $x_i - \bar{x} = x_i - y_1 = y_i$ for $i = 2, \ldots, n$. (Remember that $y_1 = \bar{x}$.) Thus

$$\sum_{i=1}^{n}(x_i - \bar{x})^2 = (-y_2 - \cdots - y_n)^2 + \sum_{i=2}^{n} y_i^2 \tag{4}$$

Now

$$f_{Y_1 \cdots Y_n}(y_1, \ldots, y_n) = n f_{X_1 \cdots X_n}(x_1, \ldots, x_n).$$

By (3) and (4), the right side becomes, in terms of the y_i's,

$$n\left(\frac{1}{\sqrt{2\pi}\sigma}\right)^n \exp\left[\frac{1}{2\sigma^2}\left(\left(-\sum_{i=2}^{n} y_i\right)^2 - \sum_{i=2}^{n} y_i^2 - n(y_1 - \mu)^2\right)\right].$$

The joint density of $Y_1, \ldots, Y_n$ is a function of y_1 times a function of $(y_2, \ldots, y_n)$, so Y_1 and $(Y_2, \ldots, Y_n)$ are independent. Since $\overline{X} = Y_1$ and [by (4)] S^2 is a function of $(Y_2, \ldots, Y_n)$,

$$\boxed{\overline{X} \text{ and } S^2 \text{ are independent}}$$

Dividing Equation (3) by σ^2 we have

$$\sum_{i=1}^{n}\left(\frac{X_i - \mu}{\sigma}\right)^2 = \frac{nS^2}{\sigma^2} + \left(\frac{\overline{X} - \mu}{\sigma/\sqrt{n}}\right)^2.$$

But $(X_i - \mu)/\sigma$ is normal (0,1) and

$$\boxed{\frac{\overline{X} - \mu}{\sigma/\sqrt{n}} \text{ is normal (0,1)}}$$

so $\chi^2(n) = (nS^2/\sigma^2) + \chi^2(1)$ with the two random variables on the right independent. If $M(t)$ is the moment-generating function of nS^2/σ^2, then $(1-2t)^{-n/2} = M(t)(1-2t)^{-1/2}$. Therefore $M(t) = (1 - 2t)^{-(n-1)/2}$, i.e.,

$$\boxed{\frac{nS^2}{\sigma^2} \text{ is } \chi^2(n-1)}$$

The random variable

$$T = \frac{\overline{X} - \mu}{S/\sqrt{n-1}}$$

is useful in situations where μ is to be estimated but the true variance σ^2 is unknown. It turns out that T has a "T distribution", which we study in the next lecture.

4.3 Performance of Various Estimates

Let S^2 be the sample variance of iid normal (μ, σ^2) random variables $X_1, \ldots, X_n$. We will look at estimates of σ^2 of the form cS^2, where c is a constant. Once again employing the centralizing technique, we write

$$E[(cS^2 - \sigma^2)^2] = E[(cS^2 - cE(S^2) + cE(S^2) - \sigma^2)^2]$$

which simplifies to

$$c^2 \operatorname{Var} S^2 + (cE(S^2) - \sigma^2)^2.$$

Since nS^2/σ^2 is $\chi^2(n-1)$, which has variance $2(n-1)$, we have $n^2(\text{Var } S^2)/\sigma^4 = 2(n-1)$. Also $nE(S^2)/\sigma^2$ is the mean of $\chi^2(n-1)$, which is $n-1$. (Or we can recall from (4.1) that $E(S^2) = (n-1)\sigma^2/n$.) Thus the mean square error is

$$\frac{c^2 2\sigma^4(n-1)}{n^2} + \left(c\frac{(n-1)}{n}\sigma^2 - \sigma^2\right)^2.$$

We can drop the σ^4 and use n^2 as a common denominator, which can also be dropped. We are then trying to minimize

$$c^2 2(n-1) + c^2(n-1)^2 - 2c(n-1)n + n^2.$$

Differentiate with respect to c and set the result equal to zero:

$$4c(n-1) + 2c(n-1)^2 - 2(n-1)n = 0.$$

Dividing by $2(n-1)$, we have $2c + c(n-1) - n = 0$, so $c = n/(n+1)$. Thus the best estimate of the form cS^2 is

$$\frac{1}{n+1}\sum_{i=1}^{n}(X_i - \overline{X})^2.$$

If we use S^2 then $c = 1$. If we us the unbiased version then $c = n/(n-1)$. Since $[n/(n+1)] < 1 < [n/(n-1)]$ and a quadratic function decreases as we move toward its minimum, w see that the biased estimate S^2 is better than the unbiased estimate $nS^2/(n-1)$, but neither is optimal under the minimum mean square error criterion. Explicitly, when $c = n/(n-1)$ we get a mean square error of $2\sigma^4/(n-1)$ and when $c = 1$ we get

$$\frac{\sigma^4}{n^2}[2(n-1) + (n-1-n)^2] = \frac{(2n-1)\sigma^4}{n^2}$$

which is always smaller, because $[(2n-1)/n^2] < 2/(n-1)$ iff $2n^2 > 2n^2 - 3n + 1$ iff $3n > 1$, which is true for every positive integer n.

For large n all these estimates are good and the difference between their performance is small.

Problems

1. Let $X_1, \ldots, X_n$ be iid, each normal (μ, σ^2), and let $\overline{X}$ be the sample mean. If c is a constant, we wish to make n large enough so that $P\{\mu - c < \overline{X} < \mu + c\} \geq .954$. Find the minimum value of n in terms of σ^2 and c. (It is independent of μ.)

2. Let $X_1, \ldots, X_{n_1}, Y_1, \ldots Y_{n_2}$ be independent random variables, with the X_i normal (μ_1, σ_1^2) and the Y_i normal (μ_2, σ_2^2). If $\overline{X}$ is the sample mean of the X_i and $\overline{Y}$ is the sample mean of the Y_i, explain how to compute the probability that $\overline{X} > \overline{Y}$.

3. Let $X_1, \ldots, X_n$ be iid, each normal (μ, σ^2), and let S^2 be the sample variance. Explain how to compute $P\{a < S^2 < b\}$.

4. Let S^2 be the sample variance of iid normal (μ, σ^2) random variables $X_i, i = 1 \ldots, n$. Calculate the moment-generating function of S^2 and from this, deduce that S^2 has a gamma distribution.

Lecture 5. The T and F Distributions

5.1 Definition and Discussion

The *T distribution* is defined as follows. Let X_1 and X_2 be independent, with X_1 normal (0,1) and X_2 chi-square with r degrees of freedom. The random variable $Y_1 = \sqrt{r}X_1/\sqrt{X_2}$ has the T distribution with r degrees of freedom.

To find the density of Y_1, let $Y_2 = X_2$. Then $X_1 = Y_1\sqrt{Y_2}/\sqrt{r}$ and $X_2 = Y_2$. The transformation is one-to-one with $-\infty < X_1 < \infty, X_2 > 0 \iff -\infty < Y_1 < \infty, Y_2 > 0$. The Jacobian is given by

$$\frac{\partial(x_1, x_2)}{\partial(y_1, y_2)} = \begin{vmatrix} \sqrt{y_2/r} & y_1/(2\sqrt{ry_2}) \\ 0 & 1 \end{vmatrix} = \sqrt{y_2/r}.$$

Thus $f_{Y_1 Y_2}(y_1, y_2) = f_{X_1 X_2}(x_1, x_2)\sqrt{y_2/r}$, which upon substitution for x_1 and x_2 becomes

$$\frac{1}{\sqrt{2\pi}} \exp[-y_1^2 y_2/2r] \frac{1}{\Gamma(r/2)2^{r/2}} y_2^{(r/2)-1} e^{-y_2/2} \sqrt{y_2/r}.$$

The density of Y_1 is

$$\frac{1}{\sqrt{2\pi}\Gamma(r/2)2^{r/2}} \int_0^\infty y_2^{[(r+1)/2]-1} \exp[-(1 + (y_1^2/r))y_2/2]\, dy_2/\sqrt{r}.$$

With $z = (1 + (y_1^2/r))y_2/2$ and the observation that all factors of 2 cancel, this becomes (with y_1 replaced by t)

$$\frac{\Gamma((r+1)/2)}{\sqrt{r\pi}\Gamma(r/2)} \frac{1}{(1 + (t^2/r))^{(r+1)/2}}, \quad -\infty < t < \infty,$$

the *T density* with r degrees of freedom.

In sampling from a normal population, $(\overline{X} - \mu)/(\sigma/\sqrt{n})$ is normal (0,1), and nS^2/σ^2 is $\chi^2(n-1)$. Thus

$$\sqrt{n-1}\frac{(\overline{X} - \mu)}{\sigma/\sqrt{n}} \quad \text{divided by} \quad \sqrt{n}S/\sigma \quad \text{is} \quad T(n-1).$$

Since σ and $\sqrt{n}$ disappear after cancellation, we have

$$\boxed{\frac{\overline{X} - \mu}{S/\sqrt{n-1}} \quad \text{is} \quad T(n-1)}$$

Advocates of defining the sample variance with $n-1$ in the denominator point out that one can simply replace σ by S in $(\overline{X} - \mu)/(\sigma/\sqrt{n})$ to get the T statistic.

Intuitively, we expect that for large n, $(\overline{X} - \mu)/(S/\sqrt{n-1})$ has approximately the same distribution as $(\overline{X} - \mu)/(\sigma/\sqrt{n})$, i.e., normal (0,1). This is in fact true, as suggested by the following computation:

$$\left(1 + \frac{t^2}{r}\right)^{(r+1)/2} = \sqrt{\left(1 + \frac{t^2}{r}\right)^r} \left(1 + \frac{t^2}{r}\right)^{1/2} \to \sqrt{e^{t^2}} \times 1 = e^{t^2/2}$$

as $r \to \infty$.

21

5.2 A Preliminary Calculation

Before turning to the F distribution, we calculate the density of $U = X_1/X_2$ where X_1 and X_2 are independent, positive random variables. Let $Y = X_2$, so that $X_1 = UY, X_2 = Y$ (X_1, X_2, U, Y) are all greater than zero). The Jacobian is

$$\frac{\partial(x_1, x_2)}{\partial(u, y)} = \begin{vmatrix} y & u \\ 0 & 1 \end{vmatrix} = y.$$

Thus $f_{UY}(u, y) = f_{X_1 X_2}(x_1, x_2)y = f_{X_1}(uy)f_{X_2}(y)$, and the density of U is

$$h(u) = \int_0^\infty y f_{X_1}(uy) f_{X_2}(y)\, dy.$$

Now we take X_1 to be $\chi^2(m)$, and X_2 to be $\chi^2(n)$. The density of X_1/X_2 is

$$h(u) = \frac{1}{2^{(m+n)/2}\Gamma(m/2)\Gamma(n/2)} u^{(m/2)-1} \int_0^\infty y^{[(m+n)/2]-1} e^{-y(1+u)/2}\, dy.$$

The substitution $z = y(1+u)/2$ gives

$$h(u) = \frac{1}{2^{(m+n)/2}\Gamma(m/2)\Gamma(n/2)} u^{(m/2)-1} \int_0^\infty \frac{z^{[(m+n)/2]-1}}{[(1+u)/2]^{[(m+n)/2]-1}} e^{-z} \frac{2}{1+u}\, dz.$$

We abbreviate $\Gamma(a)\Gamma(b)/\Gamma(a+b)$ by $\beta(a, b)$. (We will have much more to say about this when we discuss the beta distribution later in the lecture.) The above formula simplifies to

$$h(u) = \frac{1}{\beta(m/2, n/2)} \frac{u^{(m/2)-1}}{(1+u)^{(m+n)/2}}, \quad u \geq 0.$$

5.3 Definition and Discussion

The F *density* is defined as follows. Let X_1 and X_2 be independent, with $X_1 = \chi^2(m)$ and $X_2 = \chi^2(n)$. With U as in (5.2), let

$$W = \frac{X_1/m}{X_2/n} = \frac{n}{m}U$$

so that

$$f_W(w) = f_U(u) \left| \frac{du}{dw} \right| = \frac{m}{n} f_U\left(\frac{m}{n}w\right).$$

Thus W has density

$$\frac{(m/n)^{m/2}}{\beta(m/2, n/2)} \frac{w^{(m/2)-1}}{[1+(m/n)w]^{(m+n)/2}}, \quad w \geq 0,$$

the F *density* with m and n degrees of freedom.

5.4 Definitions and Calculations

The *beta function* is given by

$$\beta(a,b) = \int_0^1 x^{a-1}(1-x)^{b-1}\,dx, \quad a,b > 0.$$

We will show that

$$\beta(a,b) = \frac{\Gamma(a)\Gamma(b)}{\Gamma(a+b)}$$

which is consistent with our use of $\beta(a,b)$ as an abbreviation in (5.2). We make the change of variable $t = x^2$ to get

$$\Gamma(a) = \int_0^\infty t^{a-1}e^{-t}\,dt = 2\int_0^\infty x^{2a-1}e^{-x^2}\,dx.$$

We now use the familiar trick of writing $\Gamma(a)\Gamma(b)$ as a double integral and switching to polar coordinates. Thus

$$\Gamma(a)\Gamma(b) = 4\int_0^\infty \int_0^\infty x^{2a-1}y^{2b-1}e^{-(x^2+y^2)}\,dx\,dy$$

$$= 4\int_0^{\pi/2}d\theta \int_0^\infty (\cos\theta)^{2a-1}(\sin\theta)^{2b-1}e^{-r^2}r^{2a+2b-1}\,dr.$$

The change of variable $u = r^2$ yields

$$\int_0^\infty r^{2a+2b-1}e^{-r^2}\,dr = (1/2)\int_0^\infty u^{a+b-1}e^{-u}\,du = \Gamma(a+b)/2.$$

Thus

$$\frac{\Gamma(a)\Gamma(b)}{2\Gamma(a+b)} = \int_0^{\pi/2}(\cos\theta)^{2a-1}(\sin\theta)^{2b-1}\,d\theta.$$

Let $z = \cos^2\theta, 1 - z = \sin^2\theta, dz = -2\cos\theta\sin\theta\,d\theta = -2z^{1/2}(1-z)^{1/2}\,dz$. The above integral becomes

$$-\frac{1}{2}\int_1^0 z^{a-1}(1-z)^{b-1}\,dz = \frac{1}{2}\int_0^1 z^{a-1}(1-z)^{b-1}\,dz = \frac{1}{2}\beta(a,b)$$

as claimed. The *beta density* is

$$f(x) = \frac{1}{\beta(a,b)}x^{a-1}(1-x)^{b-1}, \quad 0 \le x \le 1 \quad (a,b > 0).$$

Problems

1. Let X have the beta distribution with parameters a and b. Find the mean and variance of X.

2. Let T have the T distribution with 15 degrees of freedom. Find the value of c which makes $P\{-c \leq T \leq c\} = .95$.

3. Let W have the F distribution with m and n degrees of freedom (abbreviated $W = F(m, n)$). Find the distribution of $1/W$.

4. A typical table of the F distribution gives values of $P\{W \leq c\}$ for $c = .9, .95, .975$ and $.99$. Explain how to find $P\{W \leq c\}$ for $c = .1, .05, .025$ and $.01$. (Use the result of Problem 3.)

5. Let X have the T distribution with n degrees of freedom (abbreviated $X = T(n)$). Show that $T^2(n) = F(1, n)$, in other words, T^2 has an F distribution with 1 and n degrees of freedom.

6. If X has the exponential density $e^{-x}, x \geq 0$, show that $2X$ is $\chi^2(2)$. Deduce that the quotient of two exponential random variables is $F(2, 2)$.

Lecture 6. Order Statistics

6.1 The Multinomial Formula

Suppose we pick a letter from $\{A, B, C\}$, with $P(A) = p_1 = .3, P(B) = p_2 = .5, P(C) = p_3 = .2$. If we do this independently 10 times, we will find the probability that the resulting sequence contains exactly 4 A's, 3 B's and 3 C's.

The probability of $AAAABBBCCC$, in that order, is $p_1^4 p_2^3 p_3^3$. To generate all favorable cases, select 4 positions out of 10 for the A's, then 3 positions out of the remaining 6 for the B's. The positions for the C's are then determined. One possibility is $BCAABACCAB$. The number of favorable cases is

$$\binom{10}{4}\binom{6}{3} = \frac{10!}{4!6!}\frac{6!}{3!3!} = \frac{10!}{4!3!3!}.$$

Therefore the probability of exactly 4 A's, 3 B's and 3 C's is

$$\frac{10!}{4!3!3!}(.3)^4(.5)^3(.2)^3$$

In general, consider n independent trials such that on each trial, the result is exactly one of the events $A_1, \ldots, A_r$, with probabilities $p_1, \ldots, p_r$ respectively. Then the probability that A_1 occurs exactly n_1 times, $\ldots$, A_r occurs exactly n_r times, is

$$p_1^{n_1} \cdots p_r^{n_r} \binom{n}{n_1}\binom{n-n_1}{n_2}\binom{n-n_1-n_2}{n_3} \cdots \binom{n-n_1-\cdots-n_{r-2}}{n_{r-1}}\binom{n_4}{n_r}$$

which reduces to the *multinomial formula*

$$\frac{n!}{n_1!\cdots n_r!}p_1^{n_1} \cdots p_r^{n_r}$$

where the p_i are nonnegative real numbers that sum to 1, and the n_i are nonnegative integers that sum to n.

Now let $X_1, \ldots, X_n$ be iid, each with density $f(x)$ and distribution function $F(x)$. Let $Y_1 < Y_2 < \cdots < Y_n$ be the X_i's arranged in increasing order, so that Y_k is the k-th smallest. In particular, $Y_1 = \min X_i$ and $Y_n = \max X_i$. The Y_k's are called the *order statistics* of the X_i's

The distributions of Y_1 and Y_n can be computed without developing any new machinery. The probability that $Y_n \leq x$ is the probability that $X_i \leq x$ for all i, which is $\prod_{i=1}^n P\{X_i \leq x\}$ by independence. But $P\{X_i \leq x\}$ is $F(x)$ for all i, hence

$$F_{Y_n}(x) = [F(x)]^n \quad \text{and} \quad f_{Y_n}(x) = n[F(x)]^{n-1}f(x).$$

Similarly,

$$P\{Y_1 > x\} = \prod_{i=1}^n P\{X_i > x\} = [1 - F(x)]^n.$$

25

Therefore

$$F_{Y_1}(x) = 1 - [1 - F(x)]^n \quad \text{and} \quad f_{Y_1}(x) = n[1 - F(x)]^{n-1} f(x).$$

We compute $f_{Y_k}(x)$ by asking how it can happen that $x \le Y_k \le x + dx$ (see Figure 6.1). There must be $k - 1$ random variables less than x, one random variable between x and $x + dx$, and $n - k$ random variables greater than x. (We are taking dx so small that the probability that more one random variable falls in $[x, x + dx]$ is negligible, and $P\{X_i > x\}$ is essentially the same as $P\{X_i > x + dx\}$. Not everyone is comfortable with this reasoning, but the intuition is very strong and can be made precise.) By the multinomial formula,

$$f_{Y_k}(x)\, dx = \frac{n!}{(k-1)!1!(n-k)!} [F(x)]^{k-1} f(x)\, dx [1 - F(x)]^{n-k}$$

so

$$f_{Y_k}(x) = \frac{n!}{(k-1)!1!(n-k)!} [F(x)]^{k-1} [1 - F(x)]^{n-k} f(x).$$

Similar reasoning (see Figure 6.2) allows us to write down the joint density $f_{Y_j Y_k}(x, y)$ of Y_j and Y_k for $j < k$, namely

$$\frac{n!}{(j-1)!(k-j-1)!(n-k)!} [F(x)]^{j-1} [F(y) - F(x)]^{k-j-1} [1 - F(y)]^{n-k} f(x) f(y)$$

for $x < y$, and 0 elsewhere. [We drop the term 1! ($=1$), which we retained for emphasis in the formula for $f_{Y_k}(x)$.]

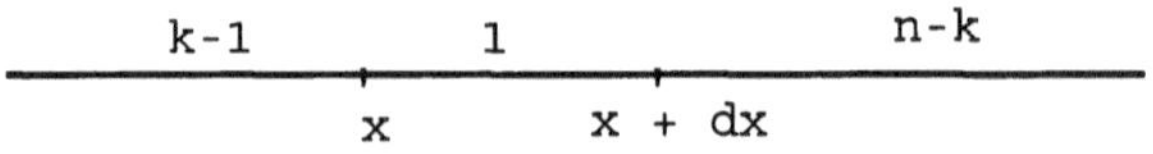

k-1	1	n-k
x	x + dx	

Figure 6.1

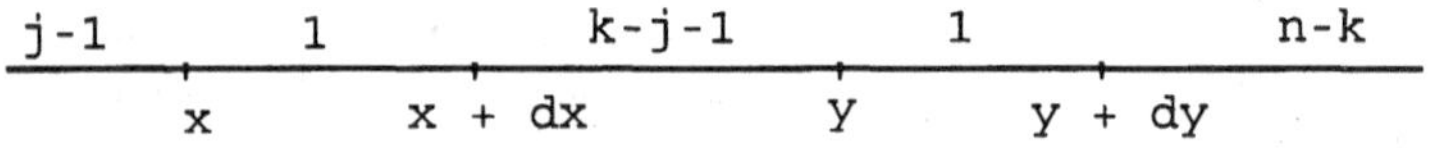

j-1	1	k-j-1	1	n-k
x	x + dx	y	y + dy	

Figure 6.2

Problems

1. Let $Y_1 < Y_2 < Y_3$ be the order statistics of X_1, X_2 and X_3, where the X_i are uniformly distributed between 0 and 1. Find the density of $Z = Y_3 - Y_1$.

2. The formulas derived in this lecture assume that we are in the continuous case (the distribution function F is continuous). The formulas do not apply if the X_i are discrete. Why not?

3. Consider order statistics where the $X_i, i = 1, \ldots, n$, are uniformly distributed between 0 and 1. Show that Y_k has a beta distribution, and express the parameters α and β in terms of k and n.

4. In Problem 3, let $0 < p < 1$, and express $P\{Y_k > p\}$ as the probability of an event associated with a sequence of n Bernoulli trials with probability of success p on a given trial. Write $P\{Y_k > p\}$ as a finite sum involving n, p and k.

Lecture 7. The Weak Law of Large Numbers

7.1 Chebyshev's Inequality

(a) If $X \geq 0$ and $a > 0$, then $P\{X \geq a\} \leq E(X)/a$.

(b) If X is an arbitrary random variable, c any real number, and $\epsilon > 0, m > 0$, then $P\{|X - c| \geq \epsilon\} \leq E(|X - c|^m)/\epsilon^m$.

(c) If X has finite mean μ and finite variance σ^2, then $P\{|X - \mu| \geq k\sigma\} \leq 1/k^2$.

This is a universal bound, but it may be quite weak in a specific cases. For example, if X is normal (μ, σ^2), abbreviated $N(\mu, \sigma^2)$, then

$$P\{|X - \mu| \geq 1.96\sigma\} = P\{|N(0,1)| \geq 1.96\} = 2(1 - \Phi(1.96)) = .05$$

where Φ is the distribution function of a normal (0,1) random variable. But the Chebyshev bound is $1/(1.96)^2 = .26$.

Proof.

(a) If X has density f, then

$$E(X) = \int_0^\infty x f(x)\, dx = \int_0^a x f(x)\, dx + \int_a^\infty x f(x)\, dx$$

so

$$E(X) \geq 0 + \int_a^\infty a f(x)\, dx = a P\{X \geq a\}.$$

(b) $P\{|X - c| \geq \epsilon\} = P\{|X - c|^m \geq \epsilon^m\} \leq E(|X - c|^m)/\epsilon^m$ by (a).

(c) By (b) with $c = \mu, \epsilon = k\sigma, m = 2$, we have

$$P\{|X - \mu| \geq k\sigma\} \leq \frac{E[(X - \mu)^2]}{k^2\sigma^2} = \frac{1}{k^2}. \quad \clubsuit$$

7.2 Weak Law of Large Numbers

Let $X_1, \ldots, X_n$ be iid with finite mean μ and finite variance σ^2. For large n, the arithmetic average of the observations is very likely to be very close to the true mean μ. Formally, if $S_n = X_1 + \cdots + X_n$, then for any $\epsilon > 0$,

$$P\{|\frac{S_n}{n} - \mu| \geq \epsilon\} \to 0 \text{ as } n \to \infty.$$

Proof.

$$P\{|\frac{S_n}{n} - \mu| \geq \epsilon\} = P\{|S_n - n\mu| \geq n\epsilon\} \leq \frac{E[(S_n - n\mu)^2]}{n^2\epsilon^2}$$

by Chebyshev (b). The term on the right is

$$\frac{\text{Var } S_n}{n^2\epsilon^2} = \frac{n\sigma^2}{n^2\epsilon^2} = \frac{\sigma^2}{n\epsilon^2} \to 0. \quad \clubsuit$$

29

7.3 Bernoulli Trials

Let $X_i = 1$ if there is a success on trial i, and $X_i = 0$ if there is a failure. Thus X_i is the indicator of a success on trial i, often written as $I[\text{Success on trial } i]$. Then S_n/n is the relative frequency of success, and for large n, this is very likely to be very close to the true probability p of success.

7.4 Definitions and Comments

The convergence illustrated by the weak law of large numbers is called *convergence in probability*. Explicitly, S_n/n converges in probability to μ. In general, $X_n \xrightarrow{P} X$ means that for every $\epsilon > 0$, $P\{|X_n - X| \geq \epsilon\} \to 0$ as $n \to \infty$. Thus for large n, X_n is very likely to be very close to X. If X_n converges in probability to X, then X_n converges to X *in distribution*: If F_n is the distribution function of X_n and F is the distribution function of X, then $F_n(x) \to F(x)$ at every x *where F is continuous*. To see that the continuity requirement is needed, look at Figure 7.1. In this example, X_n is uniformly distributed between 0 and $1/n$, and X is identically 0. We have $X_n \xrightarrow{P} 0$ because $P\{|X_n| \geq \epsilon\}$ is actually 0 for large n. However, $F_n(x) \to F(x)$ for $x \neq 0$, but not at $x = 0$.

To prove that convergence in probability implies convergence in distribution:

$$
\begin{aligned}
F_n(x) = P\{X_n \leq x\} &= P\{X_n \leq x, X > x + \epsilon\} + P\{X_n \leq x, X \leq x + \epsilon\} \\
&\leq P\{|X_n - X| \geq \epsilon\} + P\{X \leq x + \epsilon\} \\
&= P\{|X_n - X| \geq \epsilon\} + F(x + \epsilon) \\
F(x - \epsilon) = P\{X \leq x - \epsilon\} &= P\{X \leq x - \epsilon, X_n > x\} + P\{X \leq x - \epsilon, X_n \leq x\} \\
&\leq P\{|X_n - X| \geq \epsilon\} + P\{X_n \leq x\} \\
&= P\{|X_n - X| \geq \epsilon\} + F_n(x).
\end{aligned}
$$

Therefore

$$
F(x - \epsilon) - P\{|X_n - X| \geq \epsilon\} \leq F_n(x) \leq P\{|X_n - X| \geq \epsilon\} + F(x + \epsilon).
$$

Since X_n converges in probability to X, we have $P\{|X_n - X| \geq \epsilon\} \to 0$ as $n \to \infty$. If F is continuous at x, then $F(x-\epsilon)$ and $F(x+\epsilon)$ approach $F(x)$ as $\epsilon \to 0$. Thus $F_n(x)$ is boxed between two quantities that can be made arbitrarily close to $F(x)$, so $F_n(x) \to F(x)$. ♣

7.5 Some Sufficient Conditions

In practice, $P\{|X_n - X| \geq \epsilon\}$ may be difficult to compute, and it is useful to have sufficient conditions for convergence in probability that can often be easily checked.

(1) If $E[(X_n - X)^2] \to 0$ as $n \to \infty$, then $X_n \xrightarrow{P} X$.

(2) If $E(X_n) \to E(X)$ and $\text{Var}(X_n - X) \to 0$, then $X_n \xrightarrow{P} X$.

Proof. The first statement follows from Chebyshev (b):

$$
P\{|X_n - X| \geq \epsilon\} \leq \frac{E[(*X_n - X)^2]}{\epsilon^2} \to 0.
$$

To prove (2), note that

$$E[(X_n - X)^2] = \text{Var}(X_n - X) + [E(X_n) - E(X)]^2 \to 0. \quad \clubsuit$$

In this result, if X is identically equal to a constant c, then $\text{Var}(X_n - X)$ is simply $\text{Var}\, X_n$. Condition (2) then becomes $E(X_n) \to c$ and $\text{Var}\, X_n \to 0$, which implies that X_n converges in probability to c.

7.6 An Application

In normal sampling, let S_n^2 be the sample variance based on n observations. Let's show that S_n^2 is a *consistent estimate* of the true variance σ^2, that is, $S_n^2 \xrightarrow{P} \sigma^2$. Since nS_n^2/σ^2 is $\chi^2(n-1)$, we have $E(nS_n^2/\sigma^2) = (n-1)$ and $\text{Var}(nS_n^2/\sigma^2) = 2(n-1)$. Thus $E(S_n^2) = (n-1)\sigma^2/n \to \sigma^2$ and $\text{Var}(S_n^2) = 2(n-1)\sigma^4/n^2 \to 0$, and the result follows.

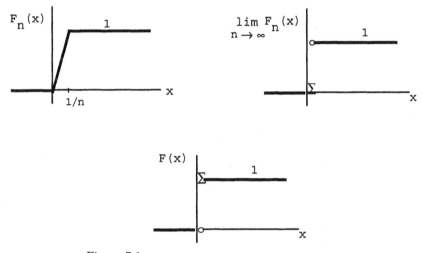

Figure 7.1

Problems

1. Let $X_1, \ldots, X_n$ be independent, not necessarily identically distributed random variables. Assume that the X_i have finite means μ_i and finite variances σ_i^2, and the variances are uniformly bounded, i.e., for some positive number M we have $\sigma_i^2 \le M$ for all i. Show that $(S_n - E(S_n))/n$ converges in probability to 0. This is a generalization of the weak law of large numbers. For if $\mu_i = \mu$ and $\sigma_i^2 = \sigma^2$ for all i, then $E(S_n) = n\mu$, so $(S_n/n) - \mu \xrightarrow{P} 0$, i.e., $S_n/n \xrightarrow{P} \mu$.

2. Toss an unbiased coin *once*. If heads, write down the sequence 10101010 $\ldots$, and if tails, write down the sequence 01010101 $\ldots$. If X_n is the n-th term of the sequence and $X = X_1$, show that X_n converges to X in distribution but not in probability.

3. Let $X_1, \ldots, X_n$ be iid with finite mean μ and finite variance σ^2. Let $\overline{X}_n$ be the sample mean $(X_1 + \cdots + X_n)/n$. Find the *limiting distribution* of $\overline{X}_n$, i.e., find a random variable X such that $\overline{X}_n \xrightarrow{d} X$.

4. Let X_n be uniformly distributed between n and $n+1$. Show that X_n does not have a limiting distribution. Intuitively, the probability has "run away" to infinity.

Lecture 8. The Central Limit Theorem

Intuitively, any random variable that can be regarded as the sum of a large number of small independent components is approximately normal. To formalize, we need the following result, stated without proof.

8.1 Theorem

If Y_n has moment-generating function M_n, Y has moment-generating function M, and $M_n(t) \to M(t)$ as $n \to \infty$ for all t in some open interval containing the origin, then $Y_n \xrightarrow{d} Y$.

8.2 Central Limit Theorem

Let $X_1, X_2, \ldots$ be iid, each with finite mean μ, finite variance σ^2, and moment-generating function M. Then

$$Y_n = \frac{\sum_{i=1}^n X_i - n\mu}{\sqrt{n}\sigma}$$

converges in distribution to a random variable that is normal $(0,1)$. Thus for large n, $\sum_{i=1}^n X_i$ is approximately normal.

We will give an informal sketch of the proof. The numerator of Y_n is $\sum_{i=1}^n (X_i - \mu)$, and the random variables $X_i - \mu$ are iid with mean 0 and variance σ^2. Thus we may assume without loss of generality that $\mu = 0$. We have

$$M_{Y_n}(t) = E[e^{tY_n}] = E\left[\exp\left(\frac{t}{\sqrt{n}\sigma} \sum_{i=1}^n X_i \right) \right].$$

The moment-generating function of $\sum_{i=1}^n X_i$ is $[M(t)]^n$, so

$$M_{Y_n}(t) = \left[M\left(\frac{t}{\sqrt{n}\sigma} \right) \right]^n.$$

Now if the density of the X_i is $f(x)$, then

$$M\left(\frac{t}{\sqrt{n}\sigma} \right) = \int_{-\infty}^{\infty} \exp\left(\frac{tx}{\sqrt{n}\sigma} \right) f(x)\, dx$$

$$= \int_{-\infty}^{\infty} \left[1 + \frac{tx}{\sqrt{n}\sigma} + \frac{t^2 x^2}{2! n\sigma^2} + \frac{t^3 x^3}{3! n^{3/2} \sigma^3} + \cdots \right] f(x)\, dx$$

$$= 1 + 0 + \frac{t^2}{2n} + \frac{t^3 \mu_3}{6 n^{3/2} \sigma^3} + \frac{t^4 \mu_4}{24 n^2 \sigma^4} + \cdots$$

where $\mu_k = E[(X_i)^k]$. If we neglect the terms after $t^2/2n$ we have, approximately,

$$M_{Y_n}(t) = \left(1 + \frac{t^2}{2n} \right)^n.$$

33

which approaches the normal $(0,1)$ moment-generating function $e^{t^2/2}$ as $n \to \infty$. This argument is very loose but it can be made precise by some estimates based on Taylor's formula with remainder.

We proved that if X_n converges in probability to X, then X_n convergence in distribution to X. There is a partial converse.

8.3 Theorem

If X_n converges in distribution to a *constant* c, then X_n converges in probability to X.
Proof. We estimate the probability that $|X_n - X| \geq \epsilon$, as follows.

$$P\{|X_n - X| \geq \epsilon\} = P\{X_n \geq c + \epsilon\} + P\{X_n \leq c - \epsilon\}$$

$$= 1 - P\{X_n < c + \epsilon\} + P\{X_n \leq c - \epsilon\}$$

Now $P\{X_n \leq c + (\epsilon/2)\} \leq P\{X_n < c + \epsilon\}$, so

$$P\{|X_n - c| \geq \epsilon\} \leq 1 - P\{X_n \leq c + (\epsilon/2)\} + P\{X_n \leq c - \epsilon\}$$

$$= 1 - F_n(c + (\epsilon/2)) + F_n(c - \epsilon).$$

where F_n is the distribution function of X_n. But as long as $x \neq c$, $F_n(x)$ converges to the distribution function of the constant c, so $F_n(x) \to 1$ if $x > c$, and $F_n(x) \to 0$ if $x < c$. Therefore $P\{|X_n - c| \geq \epsilon\} \to 1 - 1 + 0 = 0$ as $n \to \infty$. ♣

8.4 Remarks

If Y is binomial (n, p), the *normal approximation to the binomial* allows us to regard Y as approximately normal with mean np and variance npq (with $q = 1 - p$). According to Box, Hunter and Hunter, "Statistics for Experimenters", page 130, the approximation works well in practice if $n > 5$ and

$$\frac{1}{\sqrt{n}} \left| \sqrt{\frac{q}{p}} - \sqrt{\frac{p}{q}} \right| < .3$$

If, for example, we wish to estimate the probability that $Y = 50$ or 51 or 52, we may write this probability as $P\{49.5 < Y < 52.5\}$, and then evaluate as if Y were normal with mean np and variance $np(1 - p)$. This turns out to be slightly more accurate in practice than using $P\{50 \leq Y \leq 52\}$.

8.5 Simulation

Most computers an simulate a random variable that is uniformly distributed between 0 and 1. But what if we need a random variable with an arbitrary distribution function F? For example, how would we simulate the random variable with the distribution function of Figure 8.1? The basic idea is illustrated in Figure 8.2. If $Y = F(X)$ where X has the continuous distribution function F, then Y is uniformly distributed on $[0,1]$. (In Figure 8.2 we have, for $0 \le y \le 1$, $P\{Y \le y\} = P\{X \le x\} = F(x) = y$.)

Thus if X is uniformly distributed on $[0,1]$ and w want Y to have distribution function F, we set $X = F(Y), Y = \text{“}F^{-1}(X)\text{”}$.

In Figure 8.1 we must be more precise:

Case 1. $0 \le X \le 3$. Let $X = (3/70)Y + (15/70)$, $Y = (70X - 15)/3$.

Case 2. $.3 \le X \le .8$. Let $Y = 4$, so $P\{Y = 4\} = .5$ as required.

Case 3. $.8 \le X \le 1$. Let $X = (1/10)Y + (4/10)$, $Y = 10X - 4$.

In Figure 8.1, replace the $F(y)$-axis by an x-axis to visualize X versus Y. If $y = y_0$ corresponds to $x = x_0$ [i.e., $x_0 = F(y_0)$], then

$$P\{Y \le y_0\} = P\{X \le x_0\} = x_0 = F(y_0)$$

as desired.

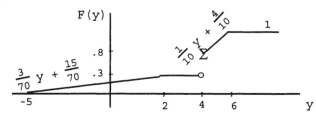

Figure 8.1

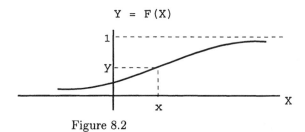

Figure 8.2

Problems

1. Let X_n be gamma (n, β), i.e., X_n has the gamma distribution with parameters n and β. Show that X_n is a sum of n independent exponential random variables, and from this derive the limiting distribution of X_n/n.

2. Show that $\chi^2(n)$ is approximately normal for large n (with mean n and variance $2n$).

3. Let $X_1, \ldots, X_n$ be iid with density f. Let Y_n be the number of observations that fall into the interval (a, b). Indicate how to use a normal approximation to calculate probabilities involving Y_n.

4. If we have 3 observations 6.45, 3.14, 4.93, and we round off to the nearest integer, we get 6, 3, 5. The sum of integers is 14, but the actual sum is 14.52. Let $X_i, i = 1, \ldots, n$ be the round-off error of the i-th observation, and assume that the X_i are iid and uniformly distributed on $(-1/2, 1/2)$. Indicate how to use a normal approximation to calculate probabilities involving the total round off error $Y_n = \sum_{i=1}^n X_i$.

5. Let $X_1, \ldots, X_n$ be iid with continuous distribution function F, and let $Y_1 < \cdots < Y_n$ be the order statistics of the X_i. Then $F(X_1), \ldots, F(X_n)$ are iid and uniformly distributed on $[0,1]$ (see the discussion of simulation), with order statistics $F(Y_1), \ldots, F(Y_n)$. Show that $n(1 - F(Y_n))$ converges in distribution to an exponential random variable.

Lecture 9. Estimation

9.1 Introduction

In effect the statistician plays a game against nature, who first chooses the "state of nature" θ (a number or k-tuple of numbers in the usual case) and performs a random experiment. We do not know θ but we are allowed to observe the value of a random variable (or random vector) X, called the *observable*, with density $f_\theta(x)$.

After observing $X = x$ we estimate θ by $\delta(x)$, which is called a *point estimate* because it produces a single number which we hope is close to θ. The main alternative is an *interval estimate* or *confidence interval*, which will be discussed in Lectures 10 and 11.

For a point estimate $\delta(x)$ to make sense physically, it must depend *only on x, not on the unknown parameter θ*. There are many possible estimates, and there are no general rules for choosing a best estimate. Some practical considerations are:

(a) How much does it cost to collect the data?

(b) Is the performance of the estimate easy to measure, for example, can we compute $P\{|\delta(x) - \theta| < \epsilon\}$?

(c) Are the advantages of the estimate appropriate for the problem at hand?

We will study several estimation methods:

1. Maximum likelihood estimates.

These estimates usually have highly desirable theoretical properties (consistency), and are frequently not difficult to compute.

2. Confidence intervals.

These estimates have a very useful practical feature. We construct an interval from the data, and we will know the probability that our (random) interval actually contains the unknown (but fixed) parameter.

3. Uniformly minimum variance unbiased estimates (UMVUE's).

Mathematical theory generates a large number of examples of these, but as we know, a biased estimate can sometimes be superior.

4. Bayes estimates.

These estimates are appropriate if it is reasonable to assume that the state of nature θ is a random variable with a known density.

In general, statistical theory produces many reasonable candidates, and practical experience will dictate the choice in a given physical situation.

9.2 Maximum Likelihood Estimates

We choose $\delta(x) = \hat{\theta}$, a value of θ that makes what we have observed as likely as possible. In other words, let $\hat{\theta}$ maximize the *likelihood function* $L(\theta) = f_\theta(x)$, with x fixed. This corresponds to basic statistical philosophy; if what we have observed is more likely under θ_2 than under θ_1, we prefer θ_2 to θ_1.

9.3 Example

Let X be binomial (n, θ). Then the probability that $X = x$ when the true parameter is θ is

$$f_\theta(x) = \binom{n}{x} \theta^x (1 - \theta)^{n-x}, x = 0, 1, \ldots, n.$$

Maximizing $f_\theta(x)$ is equivalent to maximizing $\ln f_\theta(x)$:

$$\frac{\partial}{\partial \theta} \ln f_\theta(x) = \frac{\partial}{\partial \theta} [x \ln \theta + (n - x) ln(1 - \theta)] = \frac{x}{\theta} - \frac{n - x}{1 - \theta} = 0.$$

Thus $x - \theta x - \theta n + \theta x = 0$, so $\hat{\theta} = X/n$, the relative frequency of success.

Notation: $\hat{\theta}$ will be written in terms of random variables, in this case X/n rather than x/n. Thus $\hat{\theta}$ is itself a random variable.

We have $E(\hat{\theta}) = n\theta/n = \theta$, so $\hat{\theta}$ is *unbiased*. By the weak law of large numbers, $\hat{\theta} \xrightarrow{P} \theta$, i.e., $\hat{\theta}$ is *consistent*

9.4 Example

Let $X_1, \ldots, X_n$ be iid, normal (μ, σ^2), $\theta = (\mu, \sigma^2)$. Then, with $x = (x_1, \ldots, x_n)$,

$$f_\theta(x) = \left(\frac{1}{\sqrt{2\pi}\sigma} \right)^n \exp \left[-\sum_{i=1}^n \frac{(x_i - \mu)^2}{2\sigma^2} \right]$$

and

$$\ln f_\theta(x) = -\frac{n}{2} \ln 2\pi - n \ln \sigma - \frac{1}{2\sigma^2} \sum_{i=1}^n (x_i - \mu)^2;$$

$$\frac{\partial}{\partial \mu} \ln f_\theta(x) = \frac{1}{\sigma^2} \sum_{i=1}^n (x_i - \mu) = 0, \quad \sum_{i=1}^n x_i - n\mu = 0, \quad \mu = \bar{x};$$

$$\frac{\partial}{\partial \sigma} \ln f_\theta(x) = -\frac{n}{\sigma} + \frac{1}{\sigma^3} \sum_{i=1}^n (x_i - \mu)^2 = \frac{n}{\sigma^3} \left[-\sigma^2 + \frac{1}{n} \sum_{i=1}^n (x_i - \mu)^2 \right] = 0$$

with $\mu = \bar{x}$. Thus

$$\sigma^2 = \frac{1}{n} \sum_{i=1}^n (x_i - \bar{x})^2 = s^2.$$

Case 1. μ and σ are both unknown. Then $\hat{\theta} = (\overline{X}, S^2)$.

Case 2. σ^2 is known. Then $\theta = \mu$ and $\hat{\theta} = \overline{X}$ as above. (Differentiation with respect to σ is omitted.)

Case 3. μ is known. Then $\theta = \sigma^2$ and the equation $(\partial/\partial\sigma)\ln f_\theta(x) = 0$ becomes

$$\sigma^2 = \frac{1}{n}\sum_{i=1}^{n}(x_i - \mu)^2$$

so

$$\hat{\theta} = \frac{1}{n}\sum_{i=1}^{n}(X_i - \mu)^2.$$

The sample mean $\overline{X}$ is an unbiased and (by the weak law of large numbers) consistent estimate of μ. The sample variance S^2 is a biased but consistent estimate of σ^2 (see Lectures 4 and 7).

Notation: We will abbreviate maximum likelihood estimate by MLE.

9.5 The MLE of a Function of Theta

Suppose that for a fixed x, $f_\theta(x)$ is a maximum when $\theta = \theta_0$. Then the value of θ^2 when $f_\theta(x)$ is a maximum is θ_0^2. Thus to get the MLE of θ^2, we simply square the MLE of θ. In general, if h is any function, then

$$\widehat{h(\theta)} = h(\hat{\theta}).$$

If h is continuous, then consistency is preserved, in other words:

If h is continuous and $\hat{\theta} \xrightarrow{P} \theta$, then $h(\hat{\theta}) \xrightarrow{P} h(\theta)$.

Proof. Given $\epsilon > 0$, there exists $\delta > 0$ such that if $|\hat{\theta} - \theta| < \delta$, then $|h(\hat{\theta}) - h(\theta)| < \epsilon$. Consequently,

$$P\{|h(\hat{\theta}) - h(\theta)| \geq \epsilon\} \leq P\{|\hat{\theta} - \theta| \geq \delta\} \to 0 \quad \text{as} \quad n \to \infty. \; \clubsuit$$

(To justify the above inequality, note that if the occurrence of an event A implies the occurrence of an event B, then $P(A) \leq P(B)$.)

9.6 The Method of Moments

This is sometimes a quick way to obtain reasonable estimates. We set the observed k-th moment $n^{-1}\sum_{i=1}^{n} x_i^k$ equal to the theoretical k-th moment $E(X_i^k)$ (which will depend on the unknown parameter θ). Or we set the observed k-th central moment $n^{-1}\sum_{i=1}^{n}(x_i - \mu)^k$ equal to the theoretical k-th central moment $E[(X_i - \mu)^k]$. For example, let $X_1, \ldots X_n$ be iid, gamma with $\alpha = \theta_1, \beta = \theta_2$, with $\theta_1, \theta_2 > 0$. Then $E(X_i) = \alpha\beta = \theta_1\theta_2$ and $\text{Var } X_i = \alpha\beta^2 = \theta_1\theta_2^2$ (see Lecture 3). We set

$$\overline{X} = \theta_1\theta_2, \quad S^2 = \theta_1\theta_2^2$$

and solve to get estimates θ_i^* of $\theta_i, i = 1, 2$, namely

$$\theta_2^* = \frac{S^2}{\overline{X}}, \quad \theta_1^* = \frac{\overline{X}}{\theta_2^*} = \frac{\overline{X}^2}{S^2}$$

Problems

1. In this problem, $X_1, \ldots, X_n$ are iid with density $f_\theta(x)$ or probability function $p_\theta(x)$, and you are asked to find the MLE of θ.

   (a) Poisson (θ), $\theta > 0$.

   (b) $f_\theta(x) = \theta x^{\theta - 1}, 0 < x < 1$, where $\theta > 0$. The probability is concentrated near the origin when $\theta < 1$, and near 1 when $\theta > 1$.

   (c) Exponential with parameter θ, i.e., $f_\theta(x) = (1/\theta)e^{-x/\theta}, x > 0$, where $\theta > 0$.

   (d) $f_\theta(x) = (1/2)e^{-|x-\theta|}$, where θ and x are arbitrary real numbers.

   (e) Translated exponential, i.e., $f_\theta(x) = e^{-(x-\theta)}$, where θ is an arbitrary real number and $x \geq \theta$.

2. let $X_1, \ldots, X_n$ be iid, each uniformly distributed between $\theta - (1/2)$ and $\theta + (1/2)$. Find more than one MLE of θ (so MLE's are not necessarily unique).

3. In each part of Problem 1, calculate $E(X_i)$ and derive an estimate based on the method of moments by setting the sample mean equal to the true mean. In each case, show that the estimate is consistent.

4. Let X be exponential with parameter θ, as in Problem 1(c). If $r > 0$, find the MLE of $P\{X \leq r\}$.

5. If X is binomial (n, θ) and a and b are integers with $0 \leq a \leq b \leq n$, find the MLE of $P\{a \leq X \leq b\}$.

Lecture 10. Confidence Intervals

10.1 Predicting an Election

There are two candidates A and B. If a voter is selected at random, the probability that the voter favors A is p, where p is fixed but *unknown*. We select n voters independently and ask their preference.

The number Y_n of A voters is binomial (n, p), which (for sufficiently large n), is approximately normal with $\mu = np$ and $\sigma^2 = np(1 - p)$. The relative frequency of A voters is Y_n/n. We wish to estimate the minimum value of n such that we can predict A's percentage of the vote within 1 percent, with 95 percent confidence. Thus we want

$$P\{|\frac{Y_n}{n} - p| < .01\} > .95.$$

Note that $|(Y_n/n) - p| < .01$ means that p is within .01 of Y_n/n. So this inequality can be written as

$$\frac{Y_n}{n} - .01 < p < \frac{Y_n}{n} + .01.$$

Thus the probability that the *random* interval $I_n = ((Y_n/n) - .01, (Y_n/n) + .01)$ contains the true probability p is greater than .95. We say that I_n is a *95 percent confidence interval* for p.

In general, we find confidence intervals by calculating or estimating the probability of the event that is to occur with the desired level of confidence. In this case,

$$P\left\{\left|\frac{Y_n}{n} - p\right| < .01\right\} = P\{|Y_n - np| < .01n\} = P\left\{\left|\frac{Y_n - np}{\sqrt{np(1 - p)}}\right| < \frac{.01\sqrt{n}}{\sqrt{p(1 - p)}}\right\}$$

and this is approximately

$$\Phi\left(\frac{.01\sqrt{n}}{\sqrt{p(1 - p)}}\right) - \Phi\left(\frac{-.01\sqrt{n}}{\sqrt{p(1 - p)}}\right) = 2\Phi\left(\frac{.01\sqrt{n}}{\sqrt{p(1 - p)}}\right) - 1 > .95$$

where Φ is the normal $(0,1)$ distribution function. Since $1.95/2 = .975$ and $\Phi(1.96) = .975$, we have

$$\frac{.01\sqrt{n}}{\sqrt{p(1 - p)}} > 1.96, \quad n > (196)^2 p(1 - p).$$

But (by calculus) $p(1 - p)$ is maximized when $1 - 2p = 0$, $p = 1/2$, $p(1 - p) = 1/4$. Thus $n > (196)^2/4 = (98)^2 = (100 - 2)^2 = 10000 - 400 + 4 = 9604$.

If we want to get within one tenth of one percent $(.001)$ of p with 99 percent confidence, we repeat the above analysis with .01 replaced by .001, $1.99/2 = .995$ and $\Phi(2.6) = .995$. Thus

$$\frac{.001\sqrt{n}}{\sqrt{p(1 - p)}} > 2.6, \quad n > (2600)^2/4 = (1300)^2 = 1,690,000.$$

41

To get within 3 percent with 95 percent confidence, we have

$$\frac{.03\sqrt{n}}{\sqrt{p(1-p)}} > 1.96, \quad n > \left(\frac{196}{3}\right)^2 \times \frac{1}{4} = 1067.$$

If the experiment is repeated independently a large number of times, it is very likely that our result will be within .03 of the true probability p at least 95 percent of the time. The usual statement "The margin of error of this poll is $\pm 3\%$" does not capture this idea.

Note that the accuracy of the prediction depends only on the *number of voters polled* and not in total number of votes in the population. But the model assumes sampling with replacement. (Theoretically, the same voter can be polled more than once since the voters are selected independently.) In practice, sampling is done without replacement, but if the number n of voters polled is small relative to the population size N, the error is very small.

The normal approximation to the binomial (based on the central limit theorem) is quite reliable, and is used in practice even for modest values of n; see (8.4).

10.2 Estimating the Mean of a Normal Population

Let $X_1, \ldots, X_n$ be iid, each normal (μ, σ^2). We will find a confidence interval for μ.
Case 1. The variance σ^2 is known. Then $\overline{X}$ is normal $(\mu, \sigma^2/n)$, so

$$\frac{\overline{X} - \mu}{\sigma/\sqrt{n}} \quad \text{is normal } (0,1),$$

hence

$$P\{-b < \sqrt{n}\left(\frac{\overline{X} - \mu}{\sigma}\right) < b\} = \Phi(b) - \Phi(-b) = 2\Phi(b) - 1$$

and the inequality defining the confidence interval can be written as

$$\overline{X} - \frac{b\sigma}{\sqrt{n}} < \mu < \overline{X} + \frac{b\sigma}{\sqrt{n}}.$$

We choose a symmetrical interval to minimize the length, because the normal density with zero mean is symmetric about 0. The desired confidence level determines b, which then determines the confidence interval.
Case 2. The variance σ^2 is unknown. Recall from (5.1) that

$$\frac{\overline{X} - \mu}{S/\sqrt{n-1}} \quad \text{is} \quad T(n-1)$$

hence

$$P\{-b < \frac{\overline{X} - \mu}{S/\sqrt{n-1}} < b\} = 2F_T(b) - 1$$

and the inequality defining the confidence interval can be written as

$$\overline{X} - \frac{bS}{\sqrt{n-1}} < \mu < \overline{X} + \frac{bS}{\sqrt{n-1}}.$$

10.3 A Correction Factor When Sampling Without Replacement

The following results will not be used and may be omitted, but it is interesting to measure quantitatively the effect of sampling without replacement. In the election prediction problem, let X_i be the indicator of success (i.e.,selecting an A voter) on trial i. Then $P\{X_i = 1\} = p$ and $P\{X_i = 0\} = 1 - p$. If sampling is done with replacement, then the X_i are independent and the total number $X = X_1 + \cdots + X_n$ of A voters in the sample is binomial (n, p). Thus the variance of X is $np(1-p)$. However, if sampling is done without replacement, then in effect we are drawing n balls from an urn containing N balls (where N is the size of the population), with Np balls labeled A and $N(1-p)$ labeled B. Recall from basic probability theory that

$$\text{Var } X = \sum_{i=1}^{n} \text{Var } X_i + 2 \sum_{i<j} \text{Cov}(X_i, X_j)$$

whee Cov stands for covariance. (We will prove this in a later lecture.) If $i \neq j$, then

$$E(X_i X_j) = P\{X_i = X_j = 1\} = P\{X_1 X_2 = 1\} = \frac{Np}{N} \times \frac{Np - 1}{N - 1}$$

and

$$\text{Cov}(X_i, X_j) = E(X_i X_j) - E(X_i)E(X_j) = p\left(\frac{Np - 1}{N - 1}\right) - p^2$$

which reduces to $-p(1 - p)/(N - 1)$. Now $\text{Var } X_i = p(1 - p)$, so

$$\text{Var } X = np(1 - p) - 2\binom{n}{2}\frac{p(1 - p)}{N - 1}$$

which reduces to

$$np(1 - p)\left[1 - \frac{n - 1}{N - 1}\right] = np(1 - p)\left[\frac{N - n}{N - 1}\right].$$

Thus if SE is the *standard error* (the standard deviation of X), then SE (without replacement) = SE (with replacement) times a correction factor, where the correction factor is

$$\sqrt{\frac{N - n}{N - 1}} = \sqrt{\frac{1 - (n/N)}{1 - (1/N)}}.$$

The correction factor is less than 1, and approaches 1 as $N \to \infty$, as long as $n/N \to 0$.

Note also that in sampling without replacement, the probability of getting exactly k A's in n trials is

$$\frac{\binom{Np}{k}\binom{N(1-p)}{n-k}}{\binom{N}{n}}$$

with the standard pattern $Np + N(1 - p) = N$ and $k + (n - k) = n$.

Problems

1. In the normal case [see (10.2)], assume that σ^2 is known. Explain how to compute the length of the confidence interval for μ.

2. Continuing Problem 1, assume that σ^2 is unknown. Explain how to compute the length of the confidence interval for μ, in terms of the sample standard deviation S.

3. Continuing Problem 2, explain how to compute the expected length of the confidence interval for μ, in terms of the unknown standard deviation σ. (Note that when σ is unknown, we expect a larger interval since we have less information.)

4. Let $X_1, \ldots, X_n$ be iid, each gamma with parameters α and β. If α is known, explain how to compute a confidence interval for the mean $\mu = \alpha\beta$.

5. In the binomial case [see (10.1)], suppose we specify the level of confidence and the length of the confidence interval. Explain how to compute the minimum value of n.

Lecture 11. More Confidence Intervals

11.1 Differences of Means

Let $X_1, \ldots, X_n$ be iid, each normal (μ_1, σ^2), and let $Y_1, \ldots, Y_m$ be iid, each normal (μ_2, σ^2). Assume that $(X_1, \ldots X_n)$ and $Y_1, \ldots, Y_m)$ are independent. We will construct a confidence interval for $\mu_1 - \mu_2$. In practice, the interval is often used in the following way. If the interval lies entirely to the left of 0, we have reason to believe that $\mu_1 < \mu_2$.

Since $\text{Var}(\overline{X} - \overline{Y}) = \text{Var}\,\overline{X} + \text{Var}\,\overline{Y} = (\sigma^2/n) + (\sigma^2/m)$,

$$\frac{\overline{X} - \overline{Y} - (\mu_1 - \mu_2)}{\sigma\sqrt{\frac{1}{n} + \frac{1}{m}}} \quad \text{is normal (0,1).}$$

Also, nS_1^2/σ^2 is $\chi^2(n-1)$ and mS_2^2/σ^2 is $\chi^2(m-1)$. But $\chi^2(r)$ is the sum of r independent, normal $(0,1)$ random variables, so

$$\frac{nS_1^2}{\sigma^2} + \frac{mS_2^2}{\sigma^2} \quad \text{is} \quad \chi^2(n+m-2).$$

Thus if

$$R = \sqrt{\left(\frac{nS_1^2 + mS_2^2}{n + m - 2}\right)\left(\frac{1}{n} + \frac{1}{m}\right)}$$

then

$$T = \frac{\overline{X} - \overline{Y} - (\mu_1 - \mu_2)}{R} \quad \text{is} \quad T(n+m-2).$$

Our assumption that both populations have the same variance is crucial, because the *unknown* variance can be cancelled.

If $P\{-b < T < b\} = .95$ we get a 95 percent confidence interval for $\mu_1 - \mu_2$:

$$-b < \frac{\overline{X} - \overline{Y} - (\mu_1 - \mu_2)}{R} < b$$

or

$$(\overline{X} - \overline{Y}) - bR < \mu_1 - \mu_2 < (\overline{X} - \overline{Y}) + bR.$$

If the variances σ_1^2 and σ_2^2 are *known* but possibly *unequal*, then

$$\frac{\overline{X} - \overline{Y} - (\mu_1 - \mu_2)}{\sqrt{\frac{\sigma_1^2}{n} + \frac{\sigma_2^2}{m}}}$$

is normal $(0,1)$. If R_0 is the denominator of the above fraction, we can get a 95 percent confidence interval as before: $\Phi(b) - \Phi(-b) = 2\Phi(b) - 1 > .95$,

$$(\overline{X} - \overline{Y}) - bR_0 < \mu_1 - \mu_2 < (\overline{X} - \overline{Y}) + bR_0.$$

11.2 Example

Let Y_1 and Y_2 be binomial (n_1, p_1) and (n_2, p_2) respectively. Then

$$Y_1 = X_1 + \cdots + X_{n_1} \quad \text{and} \quad Y_2 = Z_1 + \cdots + Z_{n_2}$$

where the X_i and Z_j are indicators of success on trials i and j respectively. Assume that $X_1, \ldots X_{n_1}, Z_1, \ldots, Z_{n_2}$ are independent. Now $E(Y_1/n_1) = p_1$ and $\text{Var}(Y_1/n_1) = n_1 p_1 (1 - p_1)/n_1^2 = p_1(1 - p_1)/n_1$, with similar formulas for Y_2/n_2. Thus for large n,

$$\left(\frac{Y_1}{n_1} - \frac{Y_2}{n_2} \right) - (p_1 - p_2)$$

divided by

$$\sqrt{\frac{p_1(1 - p_1)}{n_1} + \frac{p_2(1 - p_2)}{n_2}}$$

is approximately normal (0,1). But this expression cannot be used to construct confidence intervals for $p_1 - p_2$ because the denominator involves the *unknown* quantities p_1 and p_2. However, Y_1/n_1 converges in probability to p_1 and Y_2/n_2 converges in probability to p_2, and this justifies replacing p_1 by Y_1/n_1 and p_2 by Y_2/n_2 in the denominator.

11.3 The Variance

We will construct confidence intervals for the variance of a normal population. Let $X_1, \ldots, X_n$ be iid, each normal (μ, σ^2), so that nS^2/σ^2 is $\chi^2(n-1)$. If h_{n-1} is the $\chi^2(n-1)$ density and a and b are chosen to that $\int_a^b h_{n-1}(x)\,dx = 1 - \alpha$, then

$$P\{a < \frac{nS^2}{\sigma^2} < b\} = 1 - \alpha.$$

But $a < (nS^2)/\sigma^2 < b$ is equivalent to

$$\frac{nS^2}{b} < \sigma^2 < \frac{nS^2}{a}$$

so we have a confidence interval for σ^2 at confidence level $1 - \alpha$. In practice, a and b are chosen so that $\int_b^\infty h_{n-1}(x)\,dx = \int_{-\infty}^a h_{n-1}(x)\,dx$. For example, if H_{n-1} is the $\chi^2(n-1)$ distribution function and the confidence level is 95 percent, we take $H_{n-1}(a) = .025$ and $H_{n-1}(b) = 1 - .025 = .975$. This is optimal (the length of the confidence interval is minimized) when the density is symmetric about zero, and in the symmetric case we would have $a = -b$. In the nonsymmetric case (as we have here), the error is usually small.

In this example, μ is unknown. If the mean is known, we can make use of this knowledge to improve performance. Note that

$$\sum_{i=1}^n \left(\frac{X_i - \mu}{\sigma} \right)^2 \quad \text{is} \quad \chi^2(n)$$

so if

$$W = \sum_{i=1}^{n}(X_i - \mu)^2$$

and we choose a and b so that $\int_a^b h_n(x)\, dx = 1 - \alpha$, then $P\{a < (W/\sigma^2) < b\} = 1 - \alpha$. The inequality defining the confidence interval can be written as

$$\frac{W}{b} < \sigma^2 < \frac{W}{a}.$$

11.4 Ratios of Variances

Here we see an application of the F distribution. Let $X_1, \ldots, X_{n_1}$ be iid, each normal (μ_1, σ_1^2), and let $Y_1, \ldots, Y_{n_2}$ be iid, each normal (μ_2, σ_2^2). Assume that $(X_1, \ldots, X_{n_1})$ and $(Y_1, \ldots, Y_{n_2})$ are independent. Then $n_i S_i^2/\sigma_i^2$ is $\chi^2(n_i - 1), i = 1, 2$. Thus

$$\frac{(n_2 S_2^2/\sigma_2^2)/(n_2 - 1)}{(n_1 S_1^2/\sigma_1^2)/(n_1 - 1)} \quad \text{is} \quad F(n_2 - 1, n_1 - 1).$$

Let V^2 be the unbiased version of the sample variance, i.e.,

$$V^2 = \frac{n}{n-1}S^2 = \frac{1}{n-1}\sum_{i=1}^{n}(X_i - \overline{X})^2.$$

Then

$$\frac{V_2^2}{V_1^2}\frac{\sigma_1^2}{\sigma_2^2} \quad \text{is} \quad F(n_2 - 1, n_1 - 1)$$

and this allows construction of confidence intervals for σ_1^2/σ_2^2 in the usual way.

Problems

1. In (11.1), suppose the variances σ_1^2 and σ_2^2 are unknown and possibly unequal. Explain why the analysis of (11.1) breaks down.

2. In (11.1), again assume that the variances are unknown, but $\sigma_1^2 = c\sigma_2^2$ where c is a known positive constant. Show that confidence intervals for the difference of means can be constructed.

Lecture 12. Hypothesis Testing

12.1 Basic Terminology

In our general statistical model (Lecture 9), suppose that the set of possible values of θ is partitioned into two subsets A_0 and A_1, and the problem is to decide between the two possibilities $H_0 : \theta \in A_0$, the *null hypothesis*, and $H_1 : \theta \in A_1$, the *alternative*. Mathematically, it doesn't make any difference which possibility you call the null hypothesis, but in practice, H_0 is the "default setting". For example, $H_0 : \mu \le \mu_0$ might mean that a drug is no more effective than existing treatments, while $H_1 : \mu > \mu_0$ might mean that the drug is a significant improvement.

We observe x and make a decision via $\delta(x) = 0$ or 1. There are two types of errors. A *type 1 error* occurs if H_0 is true but $\delta(x) = 1$, in other words, we declare that H_1 is true. Thus in a type 1 error, we *reject H_0 when it is true*.

A *type 2 error* occurs if H_0 is false but $\delta(x) = 0$, i.e., we declare that H_0 is true. Thus in a type 2 error, we *accept H_0 when it is false*.

If H_0 [resp. H_1] means that a patient does not have [resp. does have] a particular disease, then a type 1 error is also called a *false positive*, and a type 2 error is also called a *false negative*.

If $\delta(x)$ is always 0, then a type 1 error can never occur, but a type 2 error will always occur. Symmetrically, if $\delta(x)$ is always 1, then there will always be a type 1 error, but never an error of type 2. Thus by ignoring the data altogether we can reduce one of the error probabilities to zero. To get *both* error probabilities to be mall, in practice we must increase the sample size.

We say that H_0 [resp. H_1] is *simple* if A_0 [resp. A_1] contains only one element, *composite* if A_0 [resp. H_1] contains more than one element. So in the case of *simple hypothesis vs. simple alternative*, we are testing $\theta = \theta_0$ vs. $\theta = \theta_1$. The standard example is to test the hypothesis that X has density f_0 vs. the alternative that X has density f_1.

12.2 Likelihood Ratio Tests

In the case of simple hypothesis vs. simple alternative, if we require that the probability of a type 1 error be at most α and try to minimize the probability of a type 2 error, the optimal test turns out to be a *likelihood ratio test (LRT)*, defined as follows. Let $L(x)$, the *likelihood ratio*, be $f_1(x)/f_0(x)$, and let λ be a constant. If $L(x) > \lambda$, reject H_0; if $L(x) < \lambda$, accept H_0; if $L(x) = \lambda$, do anything.

Intuitively, if what we have observed seems significantly more likely under H_1, we will tend to reject H_0. If H_0 or H_1 is composite, there is no general optimality result as there is in the simple vs. simple case. In this situation, we resort to *basic statistical philosophy*: If, assuming that H_0 is true, we witness a are event, we tend to reject H_0.

The statement that LRT's are optimal is the Neyman-Pearson lemma, to be proved at the end of the lecture. In many common examples (normal, Poisson, binomial, exponential, $L(x_1, \dots, x_n)$ can be expressed as a function of the sum of the observations, or equivalently as a function of the sample mean. This motivates consideration of tests based on $\sum_{i=1}^{n} X_i$ or on $\overline{X}$.

49

12.3 Example

Let $X_1, \ldots, X_n$ be iid, each normal (θ, σ^2). We will test $H_0 : \theta \leq \theta_0$ vs. $H_1 : \theta > \theta_0$. Under H_1, $\overline{X}$ will tend to be larger, so let's reject H_0 when $\overline{X} > c$. The *power function* of the test is defined by

$$K(\theta) = P_\theta\{\text{reject } H_0\},$$

the probability of rejecting the null hypothesis when the true parameter is θ. In this case,

$$P\{\overline{X} > c\} = P\left\{\frac{\overline{X} - \theta}{\sigma/\sqrt{n}} > \frac{c - \theta}{\sigma/\sqrt{n}}\right\} = 1 - \Phi\left(\frac{c - \theta}{\sigma/\sqrt{n}}\right)$$

(see Figure 12.1). Suppose that we specify the probability α of a type 1 error when $\theta = \theta_1$, and the probability β of a type 2 error when $\theta = \theta_2$. Then

$$K(\theta_1) = 1 - \Phi\left(\frac{c - \theta_1}{\sigma/\sqrt{n}}\right) = \alpha$$

and

$$K(\theta_2) = 1 - \Phi\left(\frac{c - \theta_2}{\sigma/\sqrt{n}}\right) = 1 - \beta.$$

If $\alpha, \beta, \sigma, \theta_1$ and θ_2 are known, we have two equations that can be solved for c and n.

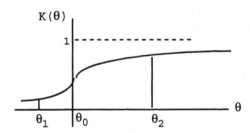

Figure 12.1

The *critical region* is the set of observations that lead to rejection. In this case, it is $\{(x_1, \ldots, x_n) : n^{-1}\sum_{i=1}^n x_i > c\}$.

The *significance level* is the largest type 1 error probability. Here it is $K(\theta_0)$, since $K(\theta)$ increases with θ.

12.4 Example

Let $H_0 : X$ is uniformly distributed on (0,1), so $f_0(x) = 1, 0 < x < 1$, and 0 elsewhere. Let $H_1 : f_1(x) = 3x^2, 0 < x < 1$, and 0 elsewhere. We take only one observation, and reject H_0 if $x > c$, where $0 < c < 1$. Then

$$K(0) = P_0\{X > c\} = 1 - c, \quad K(1) = P_1\{X > c\} = \int_c^1 3x^2\, dx = 1 - c^3.$$

If we specify the probability α of a type 1 error, then $\alpha = 1 - c$, which determines c. If β is the probability of a type 2 error, then $1 - \beta = 1 - c^3$, so $\beta = c^3$. Thus (see Figure 12.2)

$$\beta = (1 - \alpha)^3.$$

If $\alpha = .05$ then $\beta = (.95)^3 \approx .86$, which indicates that you usually can't do too well with only one observation.

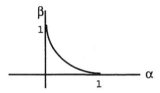

Figure 12.2

12.5 Tests Derived From Confidence Intervals

Let $X_1, \ldots, X_n$ be iid, each normal (μ_0, σ^2). In Lecture 10, we found a confidence interval for μ_0, assuming σ^2 unknown, via

$$P\left\{-b < \frac{\overline{X} - \mu_0}{S/\sqrt{n-1}} < b\right\} = 2F_T(b) - 1 \quad \text{where} \quad T = \frac{\overline{X} - \mu_0}{S/\sqrt{n-1}}$$

has the T distribution with $n - 1$ degrees of freedom.

Say $2F_T(b) - 1 = .95$, so that

$$P\left\{\left|\frac{\overline{X} - \mu_0}{S/\sqrt{n-1}}\right| \geq b\right\} = .05$$

If μ actually equals μ_0, we are witnessing an event of low probability. So it is natural to test $\mu = \mu_0$ vs. $\mu \neq \mu_0$ by rejecting if

$$\left|\frac{\overline{X} - \mu_0}{S/\sqrt{n-1}}\right| \geq b,$$

in other words, μ_0 does not belong to the confidence interval. As the true mean μ moves away from μ_0 in either direction, the probability of this event will increase, since $\overline{X} - \mu_0 = (\overline{X} - \mu) + (\mu - \mu_0)$.

Tests of $\theta = \theta_0$ vs. $\theta \neq \theta_0$ are called *two-sided*, as opposed to $\theta = \theta_0$ vs. $\theta > \theta_0$ (or $\theta = \theta_0$ vs. $\theta < \theta_0$), which are *one-sided*. In the present case, if we test $\mu = \mu_0$ vs. $\mu > \mu_0$, we reject if

$$\frac{\overline{X} - \mu_0}{S/\sqrt{n-1}} \geq b.$$

The power function $K(\mu)$ is difficult to compute for $\mu \neq \mu_0$, because $(\overline{X} - \mu_0)/(\sigma/\sqrt{n})$ no longer has mean zero. The "noncentral T distribution" becomes involved.

12.6 The Neyman-Pearson Lemma

Assume that we are testing the simple hypothesis that X has density f_0 vs. the simple alternative that X has density f_1. Let φ_λ be an LRT with parameter λ (a nonnegative constant), in other words, $\varphi_\lambda(x)$ is the probability of rejecting H_0 when x is observed, and

$$\varphi_\lambda(x) = \begin{cases} 1 & \text{if } L(x) > \lambda \\ 0 & \text{if } L(x) < \lambda \\ \text{anything} & \text{if } L(x) = \lambda \end{cases}$$

Suppose that the probability of a type 1 error using φ_λ is α_λ, and the probability of a type 2 error is β_λ. Let φ be an arbitrary test with error probabilities α and β. If $\alpha \leq \alpha_\lambda$ then $\beta \geq \beta_\lambda$. In other words, the LRT has maximum power among all tests at significance level α_λ.

Proof. We are going to assume that f_0 and f_1 are one-dimensional, but the argument works equally well when $X = (X_1, \ldots, X_n)$ and the f_i are n-dimensional joint densities. We recall from basic probability theory the *theorem of total probability*, which says that if X has density f, then for any evert A,

$$P(A) = \int_{-\infty}^{\infty} P(A|X = x) f(x)\, dx.$$

A companion theorem which we will also use later is the *theorem of total expectation*, which says that if X has density f, then for any random variable Y,

$$E(Y) = \int_{-\infty}^{\infty} E(Y|X = x) f(x)\, dx.$$

By the theorem of total probability,

$$\alpha = \int_{-\infty}^{\infty} \varphi(x) f_0(x)\, dx, \quad 1 - \beta = \int_{-\infty}^{\infty} \varphi(x) f_1(x)\, dx$$

and similarly

$$\alpha_\lambda = \int_{-\infty}^{\infty} \varphi_\lambda(x) f_0(x)\, dx, \quad 1 - \beta_\lambda = \int_{-\infty}^{\infty} \varphi_\lambda(x) f_1(x)\, dx.$$

We claim that for all x,

$$[\varphi_\lambda(x) - \varphi(x)][f_1(x) - \lambda f_0(x)] \geq 0.$$

For if $f_1(x) > \lambda f_0(x)$ then $L(x) > \lambda$, so $\varphi_\lambda(x) = 1 \geq \varphi(x)$, and if $f_1(x) < \lambda f_0(x)$ then $L(x) < \lambda$, so $\varphi_\lambda(x) = 0 \leq \varphi(x)$, proving the assertion. Now if a function is always nonnegative, its integral must be nonnegative, so

$$\int_{-\infty}^{\infty} [\varphi_\lambda(x) - \varphi(x)][f_1(x) - \lambda f_0(x)]\, dx \geq 0.$$

The terms involving f_0 translate to statements about type 1 errors, and the terms involving f_1 translate to statements about type 2 errors. Thus

$$(1 - \beta_\lambda) - (1 - \beta) - \lambda \alpha_\lambda + \lambda \alpha \geq 0,$$

which says that $\beta - \beta_\lambda \geq \lambda(\alpha_\lambda - \alpha) \geq 0$, completing the proof. ♣

12.7 Randomization

If $L(x) = \lambda$, then "do anything" means that randomization is possible, e.g., we can flip a possibly biased coin to decide whether or not to accept H_0. (This may be significant in the discrete case, where $L(x) = \lambda$ may have positive probability.) Statisticians tend to frown on this practice because two statisticians can look at exactly the same data and come to different conclusions. It is possible to adjust the significance level (by replacing "do anything" by a definite choice of either H_0 or H_1 to avoid randomization.

Problems

1. Consider the problem of testing $\theta = \theta_0$ vs. $\theta > \theta_0$, where θ is the mean of a normal population with known variance. Assume that the sample size n is fixed. Show that the test given in Example 12.3 (reject H_0 if $\overline{X} > c$) is *uniformly most powerful*. In other words, if we test $\theta = \theta_0$ vs. $\theta = \theta_1$ for any given $\theta_1 > \theta_0$, and we specify the probability α of a type 1 error, then the probability β of a type 2 error is minimized.

2. It is desired to test the null hypothesis that a die is unbiased vs. the alternative that the die is loaded, with faces 1 and 2 having probability $1/4$ and faces 3,4,5 and 6 having probability $1/8$. The die is to be tossed once. Find a most powerful test at level $\alpha = .1$, and find the type 2 error probability β.

3. We wish to test a binomial random variable X with $n = 400$ and $H_0 : p = 1/2$ vs. $H_1 : p > 1/2$. The random variable $Y = (X - np)/\sqrt{np(1-p)} = (X - 200)/10$ is approximately normal $(0,1)$, and we will reject H_0 if $Y > c$. If we specify $\alpha = .05$, then $c = 1.645$. Thus the critical region is $X > 216.45$. Suppose the actual result is $X = 220$, so that H_0 is rejected. Find the minimum value of α (sometimes called the *p-value*) for which the *given* data lead to the *opposite* conclusion (acceptance of H_0).

Lecture 13. Chi-Square Tests

13.1 Introduction

Let $X_1, \ldots, X_k$ be multinomial, i.e., X_i is the number of occurrences of the event A_i in n generalized Bernoulli trials (Lecture 6). Then

$$P\{X_1 = n_1, \ldots, X_k = n_k\} = \frac{n!}{n_1! \cdots n_k!} p_1^{n_1} \cdots p_k^{n_k}$$

where the n_i are nonnegative integers whose sum is n. Consider $k = 2$. Then X_1 is binomial (n, p_1) and $(X_1 - np_1)/\sqrt{np_1(1 - p_1)} \approx$ normal$(0,1)$. Consequently, the random variable $(X_1 - np_1)^2/np_1(1 - p_1)$ is approximately $\chi^2(1)$. But

$$\frac{(X_1 - np_1)^2}{np_1(1 - p_1)} = \frac{(X_1 - np_1)^2}{n}\left[\frac{1}{p_1} + \frac{1}{1 - p_1}\right] = \frac{(X_1 - np_1)^2}{np_1} + \frac{(X_2 - np_2)^2}{np_2}.$$

(Note that since $k = 2$ we have $p_2 = 1 - p_1$ and $X_1 - np_1 = n - X_2 - np_1 = np_2 - X_2 = -(X_2 - np_2)$, and the outer minus sign disappears when squaring.) Therefore $[(X_1 - np_1)^2/np_1] + [(X_2 - np_2)^2/np_2] \approx \chi^2(1)$. More generally, it can be shown that

$$Q = \sum_{i=1}^{k} \frac{(X_i - np_i)^2}{np_i} \approx \chi^2(k - 1).$$

where

$$\frac{(X_i - np_i)^2}{np_i} = \frac{(\text{observed frequency-expected frequency})^2}{\text{expected frequency}}.$$

We will consider three types of chi-square tests.

13.2 Goodness of Fit

We ask whether X has a specified distribution (normal, Poisson, etc.). The null hypothesis is that the multinomial probabilities are $\underline{p} = (p_1, \ldots, p_k)$, and the alternative is that $\underline{p} \neq (p_1, \ldots, p_k)$.

Suppose that $P\{\chi^2(k - 1) > c\}$ is at the desired level of significance (for example, .05). If $Q > c$ we will reject H_0. The idea is that if H_0 is in fact true, we have witnessed a rare event, so rejection is reasonable. If H_0 is false, it is reasonable to expect that some of the X_i will be far from np_i, so Q will be large.

Some practical considerations: Take n large enough so that each $np_i \geq 5$. Each time a parameter is estimated from the sample, reduced the number of degrees by 1. (A typical case: The null hypothesis is that X is Poisson (λ), but the mean λ is unknown, and is estimated by the sample mean.)

13.3 Equality of Distributions

We ask whether two or more samples come from the same underlying distribution. The observed results are displayed in a *contingency table*. This is an $h \times k$ matrix whose rows are the samples and whose columns are the attributes to be observed. For example, row i might be $(7, 11, 15, 13, 4)$, with the interpretation that in a class of 50 students taught by method of instruction i, there were 7 grades of A, 11 of B, 15 of C, 13 of D and 4 of F. The null hypothesis H_0 is that there is no difference between the various methods of instruction, i.e., $P(A)$ is the same for each group, and similarly for the probabilities of the other grades. We estimate $P(A)$ from the sample by adding all entries in column A and dividing by the total number of observations in the entire experiment. We estimate $P(B), P(C), P(D)$ and $P(F)$ in a similar fashion. The expected frequencies in row i are found by multiplying the grade probabilities by the number of entries in row i.

If there are h groups (samples), each with k attributes, then each group generates a chi-square $(k-1)$, and $k-1$ probabilities are estimated from the sample (the last probability is determined). The number of degrees of freedom is $h(k-1) - (k-1) = (h-1)(k-1)$, call it r. If $P\{\chi^2(r) > c\}$ is the desired significance level, we reject H_0 if the chi-square statistic is greater than c.

13.4 Testing For Independence

Again we have a contingency table with h rows corresponding to the possible values x_i of a random variable X, and k columns corresponding to the possible values y_j of a random variable Y. We are testing the null hypothesis that X and Y are independent.

Let R_i be the sum of the entries in row i, and let C_j be the sum of the entries in column j. Then the sum of all observations is $T = \sum_i R_i = \sum_j C_j$. We estimate $P\{X = x_i\}$ by R_i/T, and $P\{Y = y_j\}$ by C_j/T. Under the independence hypothesis H_0, $P\{X = x_i, Y = y_j\} = P\{X = x_i\}P\{Y = y_j\} = R_iC_j/T^2$. Thus the expected frequency of (x_i, y_j) is R_iC_j/T. (This gives another way to calculate the expected frequencies in (13.3). In that case, we estimated the j-th column probability by C_j/T, and multiplied by the sum of the entries in row i, namely R_i.)

In an $h \times k$ contingency table, the number of degrees of freedom is $hk - 1$ minus the number of estimated parameters:

$$hk - 1 - (h - 1 + k - 1) = hk - h - k + 1 = (h - 1)(k - 1).$$

The chi-square statistic is calculated as in (13.3). Similarly, if there are 3 attributes to be tested for independence and we form an $h \times k \times m$ contingency table, the number of degrees of freedom is

$$hkm - 1 - (h - 1) + (k - 1) + (m - 1) = hkm - h - k - m + 2.$$

Problems

1. Use a chi-square procedure to tests the null hypothesis that a random variable X has the following distribution:

$$P\{X = 1\} = .5, \quad P\{X = 2\} = .3, \quad P\{X = 3\} = .2$$

We take 100 independent observations of X, and it is observed that 1 occurs 40 times, 2 occurs 33 times, and 3 occurs 27 times. Determine whether or not we will reject the null hypothesis at significance level .05

2. Use a chi-square test to decide (at significance level .05) whether the two samples corresponding to the rows of the contingency table below came from the same underlying distribution.

	A	B	C
Sample 1	33	147	114
Sample 2	67	153	86

3. Suppose we are testing for independence in a 2×2 contingency table

$$
\begin{matrix} a & b \\ c & d \end{matrix}
$$

Show that the chi-square statistic is

$$
\frac{(ad - bc)^2(a + b + c + d)}{(a + b)(c + d)(a + c)(b + d)}
$$

(The number of degrees of freedom is $1 \times 1 = 1$.)

Lecture 14. Sufficient Statistics

14.1 Definitions and Comments

Let $X_1, \ldots, X_n$ be iid with $P\{X_i = 1\} = \theta$ and $P\{X_i = 0\} = 1 - \theta$, so $P\{X_i = x\} = \theta^x(1 - \theta)^{1-x}, x = 0, 1$. Let Y be a *statistic* for θ, i.e., a function of the observables $X_1, \ldots, X_n$. In this case we take $Y = X_1 + \cdots + X_n$, the total number of successes in n Bernoulli trials with probability θ of success on a given trial.

We claim that the conditional distribution of $X_1, \ldots, X_n$ given Y is free of θ, in other words, does not depend on θ. We say that Y is *sufficient* for θ.

To prove this, note that

$$P_\theta\{X_1 = x_1, \ldots, X_n = x_n | Y = y\} = \frac{P_\theta\{X_1 = x_1, \ldots, X_n = x_n, Y = y\}}{P_\theta\{Y = y\}}.$$

This is 0 unless $y = x_1 + \cdots + x_n$, in which case we get

$$\frac{\theta^y(1 - \theta)^{n-y}}{\binom{n}{y}\theta^y(1 - \theta)^{n-y}} = \frac{1}{\binom{n}{y}}.$$

For example, if we know that there were 3 heads in 5 tosses, the probability that the actual tosses were $HTHHT$ is $1/\binom{5}{3}$.

14.2 The Key Idea

For the purpose of making a statistical decision, we can ignore the individual random variables X_i and base the decision entirely on $X_1 + \cdots + X_n$.

Suppose that statistician A observes $X_1, \ldots, X_n$ and makes a decision. Statistician B observes $X_1 + \cdots + X_n$ only, and constructs $X_1', \ldots, X_n'$ according to the conditional distribution of $X_1, \ldots, X_n$ given Y, i.e.,

$$P\{X_1' = x_1, \ldots, X_n' = x_n | Y = y\} = \frac{1}{\binom{n}{y}}.$$

This construction is possible because the conditional distribution does not depend on the unknown parameter θ. We will show that under θ, $(X_1', \ldots, X_n')$ and $(X_1, \ldots, X_n)$ have exactly the same distribution, so anything A can do, B can do at least as well, even though B has less information.

Given $x_1, \ldots, x_n$, let $y = x_1 + \cdots + x_n$. The only way we can have $X_1' = x_1, \ldots, X_n' = x_n$ is if $Y = y$ and then B's experiment produces $X_1' = x_1, \ldots, X_n' = x_n$ given y. Thus

$$P_\theta\{X_1' = x_1, \ldots, X_n' = x_n\} = P_\theta\{Y = y\}P_\theta\{X_1' = x_1, \ldots, X_n' = x_n | Y = y\}$$

$$= \binom{n}{y}\theta^y(1 - \theta)^{n-y}\frac{1}{\binom{n}{y}} = \theta^y(1 - \theta)^{n-y} = P_\theta\{X_1 = x_1, \ldots, X_n = x_n\}.$$

59

14.3 The Factorization Theorem

Let $Y = u(X)$ be a statistic for θ; (X can be $(X_1, \ldots, X_n)$, and usually is). Then Y is sufficient for θ if and only if the density $f_\theta(x)$ of X under θ can be factored as $f_\theta(x) = g(\theta, u(x))h(x)$.

[In the Bernoulli case, $f_\theta(x_1, \ldots, x_n) = \theta^y(1-\theta)^{n-y}$ where $y = u(x) = \sum_{i=1}^{n} x_i$ and $h(x) = 1$.]

Proof. (Discrete case). If Y is sufficient, then

$$P_\theta\{X = x\} = P_\theta\{X = x, Y = u(x)\} = P_\theta\{Y = u(x)\}P\{X = x|Y = u(x)\}$$

$$= g(\theta, u(x))h(x).$$

Conversely, assume $f_\theta(x) = g(\theta, u(x))h(x)$. Then

$$P_\theta\{X = x|Y = y\} = \frac{P_\theta\{X = x, Y = y\}}{P_\theta\{Y = y\}}.$$

This is 0 unless $y = u(x)$, in which case it becomes

$$\frac{P_\theta\{X = x\}}{P_\theta\{Y = y\}} = \frac{g(\theta, u(x))h(x)}{\sum_{\{z:u(z)=y\}} g(\theta, u(z))h(z)}.$$

The g terms in both numerator and denominator are $g(\theta, y)$, which can be cancelled to obtain

$$P\{X = x|Y = y\} = \frac{h(x)}{\sum_{\{z:u(z)=y\}} h(z)}$$

which is free of θ. ♣

14.4 Example

Let $X_1, \ldots, X_n$ be iid, each normal (μ, σ^2), so that

$$f_\theta(x_1, \ldots, x_n) = \left(\frac{1}{\sqrt{2\pi}\sigma}\right)^n \exp\left[-\frac{1}{2\sigma^2}\sum_{i=1}^{n}(x_i - \mu)^2\right].$$

Take $\theta = (\mu, \sigma^2)$ and let $\bar{x} = n^{-1}\sum_{i=1}^{n} x_i$, $s^2 = n^{-1}\sum_{i=1}^{n}(x_i - \bar{x})^2$. Then

$$x_i - \bar{x} = x_i - \mu - (\bar{x} - \mu)$$

and

$$s^2 = \frac{1}{n}\left[\sum_{1}^{n}(x_i - \mu)^2 - 2(\bar{x} - \mu)\sum_{1}^{n}(x_i - \mu) + n(\bar{x} - \mu)^2\right].$$

Thus

$$s^2 = \frac{1}{n} \sum_1^n (x_i - \mu)^2 - (\bar{x} - \mu)^2.$$

The joint density is given by

$$f_\theta(x_1, \dots, x_n) = (2\pi\sigma^2)^{-n/2} e^{-ns^2/2\sigma^2} e^{-n(\bar{x}-\mu)^2/2\sigma^2}.$$

If μ and σ^2 are both unknown then $(\bar{X}, S^2)$ is sufficient (take $h(x) = 1$). If σ^2 is known then we can take $h(x) = (2\pi\sigma^2)^{-n/2} e^{-ns^2/2\sigma^2}, \theta = \mu$, and $\bar{X}$ is sufficient. If μ is known then $(h(x) = 1)$ $\theta = \sigma^2$ and $\sum_{i=1}^n (X_i - \mu)^2$ is sufficient.

Problems

In Problems 1-6, show that the given statistic $u(X) = u(X_1, \dots, X_n)$ is sufficient for θ and find appropriate functions g and h for the factorization theorem to apply.

1. The X_i are Poisson (θ) and $u(X) = X_1 + \cdots + X_n$.

2. The X_i have density $A(\theta)B(x_i), 0 < x_i < \theta$ (and 0 elsewhere), where θ is a positive real number; $u(X) = \max X_i$. As a special case, the X_i are uniformly distributed between 0 and θ, and $A(\theta) = 1/\theta, B(x_i) = 1$ on $(0, \theta)$.

3. The X_i are geometric with parameter θ, i.e., if θ is the probability of success on a given Bernoulli trial, then $P_\theta\{X_i = x\} = (1 - \theta)^x \theta$ is the probability that there will be x failures followed by the first success; $u(X) = \sum_{i=1}^n X_i$.

4. The X_i have the exponential density $(1/\theta)e^{-x/\theta}, x > 0$, and $u(X) = \sum_{i=1}^n X_i$.

5. The X_i have the beta density with parameters $a = \theta$ and $b = 2$, and $u(X) = \prod_{i=1}^n X_i$.

6. The X_i have the gamma density with parameters $\alpha = \theta$, β an arbitrary positive number, and $u(X) = \prod_{i=1}^n X_i$.

7. Show that the result in (14.2) that statistician B can do at least as well as statistician A, holds in the general case of arbitrary iid random variables X_i.

Lecture 15. Rao-Blackwell Theorem

15.1 Background From Basic Probability

To better understand the steps leading to the Rao-Blackwell theorem, consider a typical two stage experiment:

Step 1. Observe a random variable X with density $(1/2)x^2e^{-x}, x > 0$.

Step 2. If $X = x$, let Y be uniformly distributed on $(0, x)$.

Find $E(Y)$.

Method 1 via the joint density:

$$f(x,y) = f_X(x)f_Y(y|x) = \frac{1}{2}x^2e^{-x}(\frac{1}{x}) = \frac{1}{2}xe^{-x}, 0 < y < x.$$

In general, $E[g(X,Y)] = \int_{-\infty}^{\infty}\int_{-\infty}^{\infty} g(x,y)f(x,y)\,dx\,dy$. In this case, $g(x,y) = y$ and

$$E(Y) = \int_{x=0}^{\infty}\int_{y=0}^{x} y(1/2)xe^{-x}\,dy\,dx = \int_{0}^{\infty}(x^3/4)e^{-x}\,dx = \frac{3!}{4} = \frac{3}{2}.$$

Method 2 via the theorem of total expectation:

$$E(Y) = \int_{-\infty}^{\infty} f_X(x)E(Y|X = x)\,dx.$$

Method 2 works well when the conditional expectation is easy to compute. In this case it is $x/2$ by inspection. Thus

$$E(Y) = \int_{0}^{\infty} (1/2)x^2e^{-x}(x/2)\,dx = \frac{3}{2} \quad \text{as before.}$$

15.2 Comment On Notation

If, for example, it turns out that $E(Y|X = x) = x^2 + 3x + 4$, we can write $E(Y|X) = X^2 + 3X + 4$. Thus $E(Y|X)$ is a function $g(X)$ of the random variable X. When $X = x$ we have $g(x) = E(Y|X = x)$.

We now proceed to the Rao-Blackwell theorem via several preliminary lemmas.

15.3 Lemma

$E[E(X_2|X_1)] = E(X_2)$.

Proof. Let $g(X_1) = E(X_2|X_1)$. Then

$$E[g(X_1)] = \int_{-\infty}^{\infty} g(x)f_1(x)\,dx = \int_{-\infty}^{\infty} E(X_2|X_1 = x)f_1(x)\,dx = E(X_2)$$

by the theorem of total expectation. ♣

15.4 Lemma

If $\mu_i = E(X_i), i = 1, 2$, then

$$E[\{X_2 - E(X_2|X_1)\}\{E(X_2|X_1) - \mu_2\}] = 0.$$

Proof. The expectation is

$$\int_{-\infty}^{\infty} \int_{-\infty}^{\infty} [x_2 - E(X_2|X_1 = x_1)][E(X_2|X_1 = x_1) - \mu_2] f_1(x_1) f_2(x_2|x_1) \, dx_1 \, dx_2$$

$$= \int_{-\infty}^{\infty} f_1(x_1)[E(X_2|X_1 = x_1) - \mu_2] \int_{-\infty}^{\infty} [x_2 - E(X_2|X_1 = x_1)] f_2(x_2|x_1) \, dx_2 \, dx_1.$$

The inner integral (with respect to x_2) is $E(X_2|X_1 = x_1) - E(X_2|X_1 = x_1) = 0$, and the result follows. ♣

15.5 Lemma

$\operatorname{Var} X_2 \geq \operatorname{Var}[E(X_2|X_1)]$.
Proof. We have

$$\operatorname{Var} X_2 = E[(X_2 - \mu_2)^2] = E\big([\{X_2 - E(X_2|X_1\} + \{E(X_2|X_1) - \mu_2\}]^2\big)$$

$$= E[\{X_2 - E(X_2|X_1)\}^2] + E[\{E(X_2|X_1) - \mu_2\}^2] \quad \text{by (15.4)}$$

$$\geq E[\{E(X_2|X_1) - \mu_2\}^2] \quad \text{since both terms are nonnegative.}$$

But by (15.2), $E[E(X_2|X_1)] = E(X_2) = \mu_2$, so the above term is the variance of $E(X_2|X_1)$. ♣

15.6 Lemma

Equality holds in (15.5) if and only if X_2 is a function of X_1.
Proof. The argument of (15.5) shows that equality holds iff $E[\{X_2 - E(X_2|X_1)\}^2] = 0$, in other words, $X_2 = E(X_2|X_1)$. This implies that X_2 is a function of X_1. Conversely, if $X_2 = h(X_1)$, then $E(X_2|X_1) = h(X_1) = X_2$, and therefore equality holds. ♣

15.7 Rao-Blackwell Theorem

Let $X_1, \ldots, X_n$ be iid, each with density $f_\theta(x)$. Let $Y_1 = u_1(X_1, \ldots, X_n)$ be a sufficient statistic for θ, and let $Y_2 = u_2(X_1, \ldots, X_n)$ be an unbiased estimate of θ [or more generally, of a function of θ, say $r(\theta)$]. Then
(a) $\operatorname{Var}[E(Y_2|Y_1)] \leq \operatorname{Var} Y_2$, with strict inequality unless Y_2 is a function of Y_1 alone.
(b) $E[E(Y_2|Y_1)] = \theta$ [or more generally, $r(\theta)$].

Thus in searching for a minimum variance unbiased estimate of θ [or more generally, $r(\theta)$], we may restrict ourselves to functions of the sufficient statistic Y_1.
Proof. Part (a) follows from (15.5) and (15.6), and (b) follows from (15.3). ♣

15.8 Theorem

Let $Y_1 = u_1(X_1, \ldots, X_n)$ be a sufficient statistic for θ. If the maximum likelihood estimate $\hat{\theta}$ of θ is unique, then $\hat{\theta}$ is a function of Y_1.

Proof. The joint density of the X_i can be factored as

$$f_\theta(x_1, \ldots, x_n) = g(\theta, z)h(x_1, \ldots, x_n)$$

where $z = u_1(x_1, \ldots, x_n)$. Let θ_0 maximize $g(\theta, z)$. Given z, we find θ_0 by looking at all $g(\theta, z)$, so that θ_0 is a function of $u_1(X_1, \ldots, X_n) = Y_1$. But θ_0 also maximizes $f_\theta(x_1, \ldots, x_n)$, so by uniqueness, $\hat{\theta} = \theta_0$. ♣

In Lectures 15-17, we are developing methods for finding uniformly minimum variance unbiased estimates. Exercises will be deferred until Lecture 17.

Lecture 16. Lehmann-Scheffé Theorem

16.1 Definition

Suppose that Y is a sufficient statistic for θ. We say that Y is *complete* if there are no nontrivial unbiased unbiased estimates of 0 based on Y, i.e., if $E[g(Y)] = 0$ for all θ, then $P_\theta\{g(Y) = 0\} = 1$ for all θ. Thus if we have two unbiased estimates of θ based on Y, say $\varphi(Y)$ and $\psi(Y)$, then $E_\theta[\varphi(Y) - \psi(Y)] = 0$ for all θ, so that regardless of θ, $\varphi(Y)$ and $\psi(Y)$ coincide (with probability 1). So if we find one unbiased estimate of θ based on Y, we have essentially found all of them.

16.2 Theorem (Lehmann-Scheffé)

Suppose that $Y_1 = u_1(X_1, \ldots, X_n)$ is a complete sufficient statistic for θ. If $\varphi(Y_1)$ is an unbiased estimate of θ based on Y_1, then among all possible unbiased estimates of θ (whether based on Y_1 or not), $\varphi(Y_1)$ has minimum variance. We say that $\varphi(Y_1)$ is a *uniformly minimum variance unbiased estimate* of θ, abbreviated UMVUE. The term "uniformly" is used because the result holds for *all possible values* of θ.

Proof. By Rao-Blackwell, if Y_2 is any unbiased estimate of θ, then $E[Y_2|Y_1]$ is an unbiased estimate of θ with $\mathrm{Var}[E(Y_2|Y_1)] \leq \mathrm{Var}\, Y_2$. But $E(Y_2|Y_1)$ is a function of Y_1, so by completeness it must coincide with $\varphi(Y_1)$. Thus regardless of the particular value of θ, $\mathrm{Var}_\theta[\varphi(Y_1)] \leq \mathrm{Var}_\theta(Y_2)$. ♣.

Note that just as in the Rao-Blackwell theorem, the Lehmann-Scheffé result holds equally well if we are seeking a UMVUE of a function of θ. Thus we look for an unbiassed estimate of $r(\theta)$ based on the complete sufficient statistic Y_1.

16.3 Definition and Remarks

There are many situations in which complete sufficient statistics can be found quickly. The *exponential class* (or *exponential family*) consists of densities of the form

$$f_\theta(x) = a(\theta)b(x) \exp\left[\sum_{j=1}^{m} p_j(\theta)K_j(x)\right]$$

where $a(\theta) > 0, b(x) > 0, \alpha < x < \beta, \theta = (\theta_1, \ldots, \theta_k)$ with $\gamma_j < \theta_j < \delta_j, 1 \leq j \leq k$ ($\alpha, \beta, \gamma_j, \delta_j$ are constants).

There are certain regularity conditions that are assumed, but they will always be satisfied in the examples we consider, so we will omit the details. In all our examples, k and m will be equal. This is needed in the proof of completeness of the statistic to be discussed in Lecture 17. (It is not needed for sufficiency.)

16.4 Examples

1. Binomial(n, θ) where n is known. We have $f_\theta(x) = \binom{n}{x}\theta^x(1-\theta)^{n-x}, x = 0, 1, \ldots, n$, where $0 < \theta < 1$. Take $a(\theta) = (1-\theta)^n$, $b(x) = \binom{n}{x}$, $p_1(\theta) = \ln\theta - \ln(1-\theta), K_1(x) = x$. Note that $k = m = 1$.

2. Poisson(θ). The probability function is $f_\theta(x) = e^{-\theta}\theta^x/x!, x = 0, 1, \ldots$, where $\theta > 0$. We can take $a(\theta) = e^{-\theta}$, $b(x) = 1/x!$, $p_1(\theta) = \ln\theta$, $K_1(x) = x$, and $k = m = 1$.

3. Normal(μ, σ^2). The density is

$$f_\theta(x) = \frac{1}{\sqrt{2\pi}\sigma}\exp[-(x-\mu)^2/2\sigma^2], \quad -\infty < x < \infty, \quad \theta = (\mu, \sigma^2).$$

Take $a(\theta) = [1/\sqrt{2\pi}\sigma]\exp[-\mu^2/2\sigma^2]$, $b(x) = 1$, $p_1(\theta) = -1/2\sigma^2$, $K_1(x) = x^2$, $p_2(\theta) = \mu/\sigma^2$, $K_2(x) = x$, and $k = m = 2$.

4. Gamma(α, β). The density is $x^{\alpha-1}e^{-x/\beta}/[\Gamma(\alpha)\beta^\alpha]$, $x > 0$, $\theta = (\alpha, \beta)$. Take $a(\theta) = 1/[\Gamma(\alpha)\beta^\alpha]$, $b(x) = 1$, $p_1(\theta) = \alpha - 1$, $K_1(x) = \ln x$, $p_2(\theta) = -1/\beta$, $K_2(x) = x$, and $k = m = 2$.

5. Beta(a, b). The density is $[\Gamma(a+b)/\Gamma(a)\Gamma(b)]x^{a-1}(1-x)^{b-1}$, $0 < x < 1$, $\theta = (a, b)$. Take $a(\theta) = [\Gamma(a+b)/\Gamma(a)\Gamma(b)]$, $b(x) = 1$, $p_1(\theta) = a - 1$, $K_1(x) = \ln x$, $p_2(\theta) = b - 1$, $K_2(x) = \ln(1-x)$, and $k = m = 2$.

6. Negative Binomial

First we derive some properties of this distribution. In a possibly infinite sequence of Bernoulli trials, let Y_r be the number of trials required to obtain the r-th success (assume r is a known positive integer). Then $P\{Y_1 = k\}$ is the probability of $k - 1$ failures followed by a success, which is $q^{k-1}p$ where $q = 1 - p$ and $k = 1, 2, \ldots$. The moment-generating function of Y_1 is

$$M_{Y_1}(t) = E[e^{tY_1}] = \sum_{k=1}^{\infty} q^{k-1}pe^{tk}.$$

Write e^{tk} as $e^{t(k-1)}e^t$. We get

$$M_{Y_1}(t) = pe^t(1 + qe^t + (qe^t)^2 + \cdots) = \frac{pe^t}{1 - qe^t}, \quad |qe^t| < 1.$$

The random variable Y_1 is said to have the *geometric* distribution. (The slightly different random variable appearing in Problem 3 of Lecture 14 is also frequently referred to as geometric.) Now Y_r (the negative binomial random variable) is the sum of r independent random variables, each geometric, so

$$M_{Y_r}(t) = \left(\frac{pe^t}{1 - qe^t}\right)^r.$$

The event $\{Y_r = k\}$ occurs iff there are $r - 1$ successes in the first $k - 1$ trials, followed by a success on trial k. Therefore

$$P\{Y_r = k\} = \binom{k-1}{r-1}p^{r-1}q^{k-r}p, \quad x = r, r+1, r+2, \ldots.$$

We can calculate the mean and variance of Y_r from the moment-generating function, but the differentiation is not quite as messy if we introduce another random variable. Let X_r be the number of failures preceding the r-th success. Then X_r plus the number of successes preceding the r-th success is the total number of trials preceding the r-th success. Thus

$$X_r + (r-1) = Y_r - 1, \quad \text{so} \quad X_r = Y_r - r$$

and

$$M_{X_r}(t) = e^{-rt} M_{Y_r}(t) = \left(\frac{p}{1 - qe^t} \right)^r.$$

When $r = 1$ we have

$$M_{X_1}(t) = \frac{p}{1 - qe^t}, \quad E(X_1) = \frac{pqe^t}{(1 - qe^t)^2} \bigg|_{t=0} = \frac{q}{p}.$$

Since $Y_1 = X_1 + 1$ we have $E(Y_1) = 1 + (q/p) = 1/p$ and $E(Y_r) = r/p$. Differentiating the moment-generating function of X_1 again, we find that

$$E(X_1^2) = \frac{(1-q)^2 pq + pq^2 2(1-q)}{(1-q)^4} = \frac{pq(1-q)[1 - q + 2q]}{(1-q)^4} = \frac{pq(1+q)}{p^3} = \frac{q(1+q)}{p^2}.$$

Thus $\operatorname{Var} X_1 = \operatorname{Var} Y_1 = [q(1+q)/p^2] - [q^2/p^2] = q/p^2$, hence $\operatorname{Var} Y_r = rq/p^2$.

Now to show that the negative binomial distribution belongs to the exponential class:

$$P\{Y_r = x\} = \binom{x-1}{r-1} \theta^r (1-\theta)^{x-r}, \quad x = r, r+1, r+2, \dots, \theta = p.$$

Take

$$a(\theta) = \left(\frac{\theta}{1-\theta} \right)^r, \quad b(x) = \binom{x-1}{r-1}, \quad p_1(\theta) = \ln(1-\theta), \quad K_1(x) = x, \quad k = m = 1.$$

Here is the reason for the terminology "negative binomial":

$$M_{Y_r}(t) = \left(\frac{pe^t}{1 - qe^t} \right)^r = p^r e^{rt} (1 - qe^t)^{-r}.$$

To expand the moment-generating function, we use the binomial theorem with a negative exponent:

$$(1+x)^{-r} = \sum_{k=0}^{\infty} \binom{-r}{k} x^k$$

where

$$\binom{-r}{k} = \frac{-r(-r-1)\cdots(-r-k+1)}{k!}.$$

Problems are deferred to Lecture 17.

Lecture 17. Complete Sufficient Statistics For The Exponential Class

17.1 Deriving the Complete Sufficient Statistic

The density of a member of the exponential class is

$$f_\theta(x) = a(\theta)b(x) \exp\left[\sum_{j=1}^{m} p_j(\theta)K_j(x)\right]$$

so the joint density of n independent observations is

$$f_\theta(x_1, \ldots, x_n) = (a(\theta))^n \prod_{i=1}^{n} b(x_i) \exp\left[\sum_{j=1}^{m} p_j(\theta)K_j(x_1)\right] \cdots \exp\left[\sum_{j=1}^{m} p_j(\theta)K_j(x_n)\right].$$

Since $e^r e^s e^t = e^{r+s+t}$, it follows that $p_j(\theta)$ appears in the exponent multiplied by the factor $K_j(x_1) + K_j(x_2) + \cdots + K_j(x_n)$, so by the factorization theorem,

$$\left(\sum_{i=1}^{n} K_1(x_i), \ldots, \sum_{i=1}^{n} K_m(x_i)\right)$$

is sufficient for θ. This statistic is also complete. First consider $m = 1$:

$$f_\theta(x_1, \ldots, x_n) = (a(\theta))^n \prod_{i=1}^{n} b(x_i) \exp\left[p(\theta) \sum_{i=1}^{n} K(x_i)\right].$$

Let $Y_1 = \sum_{i=1}^{n} K(X_i)$; then $E_\theta[g(Y_1)]$ is given by

$$\int_{-\infty}^{\infty} \cdots \int_{-\infty}^{\infty} g\left(\sum_{i=1}^{n} K(x_i)\right) f_\theta(x_1, \ldots, x_n)\, dx_1 \cdots dx_n.$$

If $E_\theta[g(Y_1)] = 0$ for all θ, then for all θ, $g(Y_1) = 0$ with probability 1.

What we have here is analogous to a result from Laplace or Fourier transform theory: If for all t between a and b we have

$$\int_{-\infty}^{\infty} g(y)e^{ty}\, dy = 0$$

then $g = 0$. It is also analogous to the result that the moment-generating function determines the density uniquely.

When $m > 1$, the exponent in the formula for $f_\theta(x_1, \ldots, x_n)$ becomes

$$p_1(\theta) \sum_{i=1}^{n} K_1(x_i) + \cdots + p_m(\theta) \sum_{i=1}^{n} K_m(x_i)$$

71

and the argument is essentially the same as in the one-dimensional case. The transform result is as follows. If

$$\int_{-\infty}^{\infty} \cdots \int_{-\infty}^{\infty} \exp[t_1 y_1 + \cdots + t_m y_m] g(y_1, \ldots, y_n) \, dy_1 \cdots dy_m = 0$$

when $a_i < t_i < b_i, i = 1, \ldots, m$, then $g = 0$. The above integral defines a joint moment-generating function, which will appear again in connection with the multivariate normal distribution.

17.2 Example

Let $X_1, \ldots, X_n$ be iid, each normal(θ, σ^2) where σ^2 is known. The normal distribution belongs to the exponential class (see (16.4), Example 3), but in this case the term $\exp[-x^2/2\sigma^2]$ can be absorbed in $b(x)$, so only $K_2(x) = x$ is relevant. Thus $\sum_{i=1}^{n} X_i$, equivalently $\overline{X}$, is sufficient (as found in Lecture 14) and complete. Since $E(\overline{X}) = \theta$, it follows that $\overline{X}$ is a UMVUE of θ.

Let's find a UNVUE of θ^2. The natural conjecture that it is $(\overline{X})^2$ is not quite correct. Since $\overline{X} = n^{-1} \sum_{i=1}^{n} X_i$, we have $\operatorname{Var} \overline{X} = \sigma^2/n$. Thus

$$\frac{\sigma^2}{n} = E[(\overline{X})^2] - (E\overline{X})^2 = E[(\overline{X})^2] - \theta^2,$$

hence

$$E\left[(\overline{X})^2 - \frac{\sigma^2}{n}\right] = \theta^2$$

and we have an unbiased estimate of θ^2 based on the complete sufficient statistic $\overline{X}$. Therefore $(\overline{X})^2 - [\sigma^2/n]$ is a UMVUE of θ^2.

17.3 A Cautionary Tale

Restricting to unbiased estimates is not always a good idea. Let X be Poisson(θ), and take $n = 1$, i.e., only one observation is made. From (16.4), Example 2, X is a complete sufficient statistic for θ. Now

$$E[(-1)^X] = \sum_{k=0}^{\infty} (-1)^k \frac{e^{-\theta}\theta^k}{k!} = e^{-\theta} \sum_{k=0}^{\infty} \frac{(-\theta)^k}{k!} = e^{-\theta} e^{-\theta} = e^{-2\theta}.$$

thus $(-1)^X$ is a UMVUE of $e^{-2\theta}$. But $Y \equiv 1$ is certainly a better estimate, since 1 is closer to $e^{-2\theta}$ than is -1. Estimating a positive quantity $e^{-2\theta}$ by a random variable which can be negative is not sensible.

Note also that the maximum likelihood estimate of θ is X (Lecture 9, Problem 1a), so the MLE of $e^{-2\theta}$ is e^{-2X}, which looks better than Y.

Problems

1. Let X be a random variable that has zero mean for all possible values of θ. For example, X can be uniformly distributed between $-\theta$ and θ, or normal with mean 0 and variance θ. Give an example of a sufficient statistic for θ that is not complete.

2. Let $f_\theta(x) = \exp[-(x - \theta)], \theta < x < \infty$, and 0 elsewhere. Show that the first order statistic $Y_1 = \min X_i$ is a complete sufficient statistic for θ, and find a UMVUE of θ.

3. Let $f_\theta(x) = \theta x^{\theta-1}, 0 < x < 1$, where $\theta > 0$. Show that $u(X_1, \dots, X_n) = \left[\prod_{i=1}^n X_i\right]^{1/n}$ is a complete sufficient statistic for θ, and that the maximum likelihood estimate $\hat\theta$ is a function of $u(X_1, \dots, X_n)$.

4. The density $f_\theta(x) = \theta^2 x \exp[-\theta x], x > 0$, where $\theta > 0$, belongs to the exponential class, and $Y = \sum_{i=1}^n X_i$ is a complete sufficient statistic for θ. Compute the expectation of $1/Y$ under θ, and from the result find the UMVUE of θ.

5. Let Y_1 be binomial (n, θ), so that $Y_1 = \sum_{i=1}^n X_i$, where X_i is the indicator of a success on trial i. [Thus each X_i is binomial $(1, \theta)$.] By Example 1 of (16.4), the X_i, as well as Y_1, belong to the exponential class, and Y_1 is a complete sufficient statistic for θ. Since $E(Y_1) = n\theta$, Y_1/n is a UMVUE of θ.

   Let $Y_2 = (X_1 + X_2)/2$. In an effortless manner, find $E(Y_2|Y_1)$.

6. Let X be normal with mean 0 and variance θ, so that by Example 3 of (16.4), $Y = \sum_{i=1}^n X_i^2$ is a complete sufficient statistic for θ. Find the distribution of Y/θ, and from this find the UMVUE of θ^2.

7. Let $X_1, \dots, X_n$ be iid, each Poisson (θ), where $\theta > 0$. (Then $Y = \sum_{i=1}^n X_i$ is a complete sufficient statistic for θ.) Let I be the indicator of $\{X_1 \leq 1\}$.
   (a) Show that $E(I|Y)$ is the UMVUE of $P\{X_1 \leq 1\} = (1 + \theta)\exp(-\theta)$. Thus we need to evaluate $P\{X_1 = 0|Y = y\} + P\{X_1 = 1|Y = y\}$. When $y = 0$, the first term is 1 and the second term is 0.
   (b) Show that if $y > 0$, the conditional distribution of X_1 (or equally well, of any X_i) is binomial $(y, 1/n)$.
   (c) Show that

$$E(I|Y) = \left(\frac{n-1}{n}\right)^Y \left[1 + \frac{Y}{n-1}\right]$$

8. Let $\theta = (\theta_1, \theta_2)$ and $f_\theta(x) = (1/\theta_2)\exp[(x - \theta_1)/\theta_2], x > \theta_1$ (and 0 elsewhere) where θ_1 is an arbitrary real number and $\theta_2 > 0$. Show that the statistic $(\min_i X_i, \sum_{i=1}^n X_i)$ is sufficient for θ.

Lecture 18. Bayes Estimates

18.1 Basic Assumptions

Suppose we are trying to estimate the state of nature θ. We observe $X = x$, where X has density $f_\theta(x)$, and make decision $\delta(x) = $ our estimate of θ when x is observed. We incur a loss $L(\theta, \delta(x))$, assumed nonnegative. We now *assume* that θ is random with density $h(\theta)$. The Bayes solution minimizes the *Bayes risk* or *average loss*

$$B(\delta) = \int_{-\infty}^{\infty} \int_{-\infty}^{\infty} h(\theta) f_\theta(x) L(\theta, \delta(x)) \, d\theta \, dx.$$

Note that $h(\theta) f_\theta(x) = h(\theta) f(x|\theta)$ is the joint density of θ and x, which can also be expressed as $f(x) f(\theta|x)$. Thus

$$B(\delta) = \int_{-\infty}^{\infty} f(x) \left[\int_{-\infty}^{\infty} L(\theta, \delta(x)) f(\theta|x) \, d\theta \right] dx.$$

Since $f(x)$ is nonnegative, it is sufficient to minimize $\int_{-\infty}^{\infty} L(\theta, \delta(x)) f(\theta|x) \, d\theta$ for each x. The resulting δ is called the *Bayes estimate* of θ. Similarly, to estimate a function of θ, say $\gamma(\theta)$, we minimize $\int_{-\infty}^{\infty} L(\gamma(\theta), \delta(x)) f(\theta|x) \, d\theta$.

We can jettison a lot of terminology by recognizing that our problem is to observe a random variable X and estimate a random variable Y by $g(X)$. We must minimize $E[L(Y, g(X)]$.

18.2 Quadratic Loss Function

We now assume that $L(Y, g(X)) = (Y - g(X))^2$. By the theorem of total expectation,

$$E(Y - g(X))^2 = \int_{-\infty}^{\infty} E[(Y - g(X))^2 | X = x] f(x) \, dx$$

and as above, it suffices to minimize the quantity in brackets for each x. If we let $z = g(x)$, we are minimizing $z^2 - 2E(Y|X = x)z + E(Y^2|X = x)$ by choice of z. Now $Az^2 - 2Bz + C$ is a minimum when $z = B/A = E(Y|X = x)/1$, and we conclude that

$$\boxed{E[(Y - g(X))^2] \quad \text{is minimized when} \quad g(x) = E(Y|X = x).}$$

What we are doing here is minimizing $E[(W - c)^2] = c^2 - 2E(W)c + E(W^2)$ by our choice of c, and the minimum occurs when $c = E(W)$.

18.3 A Different Loss Function

Suppose that we want to minimize $E(|W - c|)$. We have

$$E(|W - c|) = \int_{-\infty}^{c} (c - w) f(w) \, dw + \int_{c}^{\infty} (w - c) f(w) \, dw$$

$$= c \int_{-\infty}^{c} f(w)\, dw - \int_{-\infty}^{c} w f(w)\, dw + \int_{c}^{\infty} w f(w)\, dw - c \int_{c}^{\infty} f(w)\, dw.$$

Differentiating with respect to c, we get

$$cf(c) + \int_{-\infty}^{c} f(w)\, dw - cf(c) - cf(c) + cf(c) - \int_{c}^{\infty} f(w)\, dw$$

which is 0 when $\int_{-\infty}^{c} f(w)\, dw = \int_{c}^{\infty} f(w)\, dw$, in other words when C is a *median* of W. Thus $E(|Y - g(X)|)$ is minimized when $g(x)$ is a median of the conditional distribution of Y given $X = x$.

18.4 Back To Quadratic Loss

In the statistical decision problem with quadratic loss, the Bayes estimate is

$$\delta(x) = E[\theta|X = x] = \int_{-\infty}^{\infty} \theta f(\theta|x)\, d\theta$$

and

$$f(\theta|x) = \frac{f(\theta, x)}{f(x)} = \frac{h(\theta) f(x|\theta)}{f(x)}.$$

Thus

$$\boxed{\delta(x) = \frac{\int_{-\infty}^{\infty} \theta h(\theta) f_\theta(x)\, d\theta}{\int_{-\infty}^{\infty} h(\theta) f_\theta(x)\, d\theta}}$$

If we are estimating a function of θ, say $\gamma(\theta)$, replace θ by $\gamma(\theta)$ in the integral in the numerator.

Problems

1. Let X be binomial(n, θ), and let the density of θ be

$$h(\theta) = \frac{\theta^{r-1}(1 - \theta)^{s-1}}{\beta(r, s)} \quad \text{[beta}(r, s)].$$

Show that the Bayes estimate with quadratic loss is

$$\delta(x) = \frac{r + x}{r + s + n}, \qquad x = 0, 1, \dots, n.$$

2. For this estimate, show that the *risk function* $R_\delta(\theta)$, defined as the average loss using δ when the parameter is θ, is

$$\frac{1}{(r + s + n)^2}[((r + s)^2 - n)\theta^2 + (n - 2r(r + s))\theta + r^2]$$

3. Show that if $r = s = \sqrt{n}/2$, then $R_\delta(\theta)$ is a constant, independent of θ.

4. Show that a Bayes estimate δ with constant risk (as in Problem 3) is *minimax*, that is, δ minimizes $\max_\theta R_\delta(\theta)$.

Lecture 19. Linear Algebra Review

19.1 Introduction

We will assume for the moment that matrices have complex numbers as entries, but the complex numbers will soon disappear. If A is a matrix, the conjugate transpose of A will be denoted by A^*. Thus if

$$A = \begin{bmatrix} a + bi & c + di \\ e + fi & g + hi \end{bmatrix} \quad \text{then} \quad A^* = \begin{bmatrix} a - bi & e - fi \\ c - di & g - hi \end{bmatrix}.$$

The transpose is

$$A' = \begin{bmatrix} a + bi & e + fi \\ c + di & g + hi \end{bmatrix}.$$

Vectors X, Y, etc., will be regarded as column vectors. The *inner product (dot product)* of n-vectors X and Y is

$$< X, Y > = x_1 \bar{y}_1 + \cdots + x_n \bar{y}_n$$

where the overbar indicates complex conjugate. Thus $< X, Y > = Y^* X$. If c is any complex number, then $< cX, Y > = c < X, Y >$ and $< X, cY > = \bar{c} < X, Y >$. The vectors X and Y are said to be *orthogonal (perpendicular)* if $< X, Y > = 0$. For an arbitrary n by n matrix B,

$$< BX, Y > = < X, B^* Y >$$

because $< X, B^* Y > = (B^* Y)^* X = Y^* B^{**} X = Y^* BX = < BX, Y >$.

Our interest is in *real symmetric matrices*, and "symmetric" will always mean "real symmetric". If A is symmetric then

$$< AX, Y > = < X, A^* Y > = < X, AY >.$$

The *eigenvalue problem* is $AX = \lambda X$, or $(A - \lambda I)X = 0$, where I is the identity matrix, i.e., the matrix with 1's down the main diagonal and 0's elsewhere. A nontrivial solution ($X \neq 0$) exists iff $\det(A - \lambda I) = 0$. In this case, λ is called an *eigenvalue* of A and a nonzero solution is called an *eigenvector*.

19.2 Theorem

If A is symmetric then A has real eigenvalues.

Proof. Suppose $AX = \lambda X$ with $X \neq 0$. then $< AX, Y > = < X, AY >$ with $Y = X$ gives $< \lambda X, X > = < X, \lambda X >$, so $(\lambda - \bar{\lambda}) < X, X > = 0$. But $< X, X > = \sum_{i=1}^{n} |x_i|^2 \neq 0$, and therefore $\lambda = \bar{\lambda}$, so λ is real. ♣

The important conclusion is that for a symmetric matrix, the eigenvalue problem can be solved using only real numbers.

19.3 Theorem

If A is symmetric, then eigenvectors of distinct eigenvalues are orthogonal.

Proof. Suppose $AX_1 = \lambda_1 X_1$ and $AX_2 = \lambda_2 X_2$. Then $< AX_1, X_2 >=< X_1, AX_2 >$, so $< \lambda_1 X_1, X_2 >=< X_1, \lambda_2 X_2 >$. Since λ_2 is real we have $(\lambda_1 - \lambda_2) < X_1, X_2 >= 0$. But we are assuming that we have two distinct eigenvalues, so that $\lambda_1 \neq \lambda_2$. Therefore we must have $< X_1, X_2 >= 0$. ♣

19.4 Orthogonal Decomposition Of Symmetric Matrices

Assume A symmetric with distinct eigenvalues $\lambda_1, \dots, \lambda_n$. The assumption that the λ_i are distinct means that the equation $\det(A - \lambda I) = 0$, a polynomial equation in λ of degree n, has no repeated roots. This assumption is actually unnecessary, but it makes the analysis much easier.

Let $AX_i = \lambda_i X_i$ with $X_i \neq 0, i = 1, \dots, n$. Normalize the eigenvectors so that $\|X_i\|$, the length of X_i, is 1 for all i. (The length of the vector $x = (x_1, \dots, x_n)$ is

$$\|x\| = \left(\sum_{i=1}^{n} |x_i|^2 \right)^{1/2}$$

hence $\|x\|^2 =< x, x >$.) Thus we have $AL = LD$, where

$$L = [X_1|X_2|\cdots|X_n|] \quad \text{and} \quad D = \begin{bmatrix} \lambda_1 & & 0 \\ & \ddots & \\ 0 & & \lambda_n \end{bmatrix}.$$

To verify this, note that multiplying L on the right by a diagonal matrix with entries $\lambda_1, \dots, \lambda_n$ multiplies column i of L (namely X_i) by λ_i. (Multiplying on the left by D would multiply row i by λ_i.) Therefore

$$LD = [\lambda_1 X_1|\lambda_2 X_2|\cdots|\lambda_n X_n|] = AL.$$

The columns of the square matrix L are mutually perpendicular unit vectors; such a matrix is said to be *orthogonal*. The transpose of L can be pictured as follows:

$$L' = \begin{bmatrix} X_1' \\ X_2' \\ \vdots \\ X_n' \end{bmatrix}$$

Consequently $L'L = I$. Since L is nonsingular ($\det I = 1 = \det L' \det L$), L has an inverse, which must be L'. to see this, multiply the equation $L'L = I$ on the right by L^{-1} to get $L'I = L^{-1}$, i.e., $L' = L^{-1}$. Thus $LL' = I$.

Since a matrix and its transpose have the same determinant, $(\det L)^2 = 1$, so the determinant of L is ± 1.

Finally, from $AL = LD$ we get

$$\boxed{L'AL = D}$$

We have shown that *every symmetric matrix (with distinct eigenvalues) can be orthogonally diagonalized.*

19.5 Application To Quadratic Forms

Consider a *quadratic form*

$$X'AX = \sum_{i,j=1}^{n} a_{i,j} x_i x_j.$$

If we change variables by $X = LY$, then

$$X'AX = Y'L'ALY = Y'DY = \sum_{i=1}^{n} \lambda_i y_i^2.$$

The symmetric matrix A is said to be *nonnegative definite* if $X'AX \geq 0$ for all X. Equivalently, $\sum_{i=1}^{n} \lambda_i y_i^2 \geq 0$ for all Y. Set $y_i = 1, y_j = 0$ for all $j \neq i$ to conclude that A is nonnegative definite if and only if *all eigenvalues of A are nonnegative.* The symmetric matrix is said to be *positive definite* if $X'AX > 0$ except when all $x_i = 0$. Equivalently, *all eigenvalues of A are strictly positive.*

19.6 Example

Consider the quadratic form

$$q = 3x^2 + 2xy + 3y^2 = (x, y) \begin{bmatrix} 3 & 1 \\ 1 & 3 \end{bmatrix} \begin{pmatrix} x \\ y \end{pmatrix}$$

Then

$$A = \begin{bmatrix} 3 & 1 \\ 1 & 3 \end{bmatrix}, \quad \det(A - \lambda I) = \begin{vmatrix} 3 - \lambda & 1 \\ 1 & 3 - \lambda \end{vmatrix} = \lambda^2 - 6\lambda + 8 = 0$$

and the eigenvalues are $\lambda = 2$ and $\lambda = 4$. When $\lambda = 2$, the equation $A(x, y)' = \lambda(x, y)'$ reduces to $x + y = 0$. Thus $(1, -1)$ is an eigenvector. Normalize it to get $(1/\sqrt{2}, -1/\sqrt{2})'$. When $\lambda = 4$ we get $-x + y = 0$ and the normalized eigenvector is $(1/\sqrt{2}, 1/\sqrt{2})'$. Consequently,

$$L = \begin{bmatrix} 1/\sqrt{2} & 1/\sqrt{2} \\ -1/\sqrt{2} & 1/\sqrt{2} \end{bmatrix}$$

and a direct matrix computation yields

$$L'AL = \begin{bmatrix} 2 & 0 \\ 0 & 4 \end{bmatrix} = D$$

as expected. If $(x, y)' = L(v, w)'$, i.e., $x = (1/\sqrt{2})v + (1/\sqrt{2})w$, $y = (-1/\sqrt{2})v + (1/\sqrt{2})w$, then

$$q = 3\left[\frac{v^2}{2} + \frac{w^2}{2} + vw\right] + 2\left[-\frac{v^2}{2} + \frac{w^2}{2}\right] + 3\left[\frac{v^2}{2} + \frac{w^2}{2} - vw\right].$$

Thus $q = 2v^2 + 4w^2 = (v, w)D(v, w)'$, as expected.

Lecture 20. Correlation

20.1 Definitions and Comments

Let X and Y be random variables with finite mean and variance. Denote the mean of X by μ_1 and the mean of Y by μ_2, and let $\sigma_1^2 = \text{Var } X$ and $\sigma_2^2 = \text{Var } Y$. Note that $E(XY)$ must be finite also, because $-X^2 - Y^2 \leq 2XY \leq X^2 + Y^2$. The *covariance* of X and Y is defined by

$$\text{Cov}(X, Y) = E[(X - \mu_1)(Y - \mu_2)]$$

and it follows that

$$\text{Cov}(X, Y) = E(XY) - \mu_1 E(Y) - \mu_2 E(X) + \mu_1 \mu_2 = E(XY) - E(X)E(Y).$$

Thus $\text{Cov}(X, Y) = \text{Cov}(Y, X)$. Since expectation is linear, we have $\text{Cov}(aX, bY) = ab\,\text{Cov}(X, Y)$, $\text{Cov}(X, Y + Z) = \text{Cov}(X, Y) + \text{Cov}(X, Z)$, $\text{Cov}(X + Y, Z) = \text{Cov}(X, Z) + \text{Cov}(Y, Z)$, and $\text{Cov}(X + a, Y + b) = \text{Cov}(X, Y)$. Also, $\text{Cov}(X, X) = E(X^2) - (EX)^2 = \text{Var } X$.

The *correlation coefficient* is a normalized covariance:

$$\rho = \frac{\text{Cov}(X, Y)}{\sigma_1 \sigma_2}.$$

The correlation coefficient is a measure of *linear dependence* between X and Y. To see this, estimate Y by $AX + b$, equivalently (to simplify the calculation) estimate $Y - \mu_2$ by $c(X - \mu_1) + d$, choosing c and d to minimize

$$E[(Y - \mu_2 - (c(X - \mu_1) + d))^2] = \sigma_2^2 - 2c\,\text{Cov}(X, Y) + c^2\sigma_1^2 + d^2.$$

Note that $E[2cd(X - \mu_1)] = 0$ since $E(X) = \mu_1$, and similarly $E[2d(Y - \mu_2)] = 0$. We can't do any better than to take $d = 0$, so we need to minimize $\sigma_2^2 - 2c\rho\sigma_1\sigma_2 + c^2\sigma_1^2$ by choice of c. Differentiating with respect to c, we have $-2\rho\sigma_1\sigma_2 + 2c\sigma_1^2$, hence

$$\boxed{c = \rho\frac{\sigma_2}{\sigma_1}}$$

The minimum expectation is

$$\sigma_2^2 - 2\rho\frac{\sigma_2}{\sigma_1}\rho\sigma_1\sigma_2 + \rho^2\frac{\sigma_2^2}{\sigma_1^2}\sigma_1^2 = \boxed{\sigma_2^2(1 - \rho^2)}$$

The expectation of a nonnegative random variable is nonnegative, so

$$\boxed{-1 \leq \rho \leq 1}$$

For a fixed σ_2, the closer $|\rho|$ is to 1, the better the estimate of Y by $aX + b$. If $|\rho| = 1$ then the minimum expectation is 0, so (with probability 1)

$$Y - \mu_2 = c(X - \mu_1) = \rho\frac{\sigma_2}{\sigma_1}(X - \mu_1) \quad \text{with} \quad \rho = \pm 1.$$

81

20.2 Theorem

If X and Y are independent then X and Y are uncorrelated ($\rho = 0$) but not conversely.
Proof. Assume X and Y are independent. Then

$$E[(X - \mu_1)(Y - \mu_2)] = E(X - \mu_1)E(Y - \mu_2) = 0.$$

For the counterexample to the converse, let $X = \cos\theta, Y = \sin\theta$, where θ is uniformly distributed on $(0, 2\pi)$. Then

$$E(X) = \frac{1}{2\pi} \int_0^{2\pi} \cos\theta \, d\theta = 0, \quad E(Y) = \frac{1}{2\pi} \int_0^{2\pi} \sin\theta \, d\theta = 0,$$

and

$$E(XY) = E[(1/2)\sin 2\theta] = \frac{1}{4\pi} \int_0^{2\pi} \sin 2\theta \, d\theta = 0,$$

so $\rho = 0$. But $X^2 + Y^2 = 1$, so X and Y are not independent. ♣

20.3 The Cauchy-Schwarz Inequality

This result, namely

$$|E(XY)|^2 \leq E(X^2)E(Y^2)$$

is closely related to $-1 \leq \rho \leq 1$. Indeed, if we replace X by $X - \mu_1$ and Y by $Y - \mu_2$, the inequality says that $[\text{Cov}(X, Y)]^2 \leq \sigma_1^2\sigma_2^2$, i.e., $(\rho\sigma_1\sigma_2)^2 \leq \sigma_1^2\sigma_2^2$, which gives $\rho^2 \leq 1$. Thus Cauchy-Schwarz implies $-1 \leq \rho \leq 1$.
Proof. Let $h(\lambda) = E[(\lambda X + Y)^2] = \lambda^2 E(X^2) + 2\lambda E(XY) + E(Y^2)$. Since $h(\lambda) \geq 0$ for all λ, the quadratic equation $h(\lambda) = 0$ has no real roots or at worst a real repeated root. Therefore the discriminant is negative or at worst 0. Thus $[2E(XY)]^2 - 4E(X^2)E(Y^2) \leq 0$, and the result follows. ♣

As a special case, let $P\{X = x_i\} = 1/n, 1 \leq i \leq n$. If $X = x_i$, take $Y = y_i$. (The x_i and y_i are arbitrary real numbers.) Then the Cauchy-Schwarz inequality becomes

$$\left(\sum_{i=1}^n x_i y_i\right)^2 \leq \left(\sum_{i=1}^n x_i^2\right)\left(\sum_{i=1}^n y_i^2\right).$$

(There will be a factor of $1/n^2$ on each side of the inequality, which will cancel.) This is the result originally proved by Cauchy. Schwarz proved the analogous formula for integrals:

$$\left(\int_a^b f(x)g(x)\, dx\right)^2 \leq \int_a^b [f(x)]^2\, dx \int_a^b [g(x)]^2\, dx.$$

Since an integral can be regarded as the limit of a sum, the integral result can be proved from the result for sums.

We know that if $X_1, \ldots, X_n$ are independent, then the variance of the sum of the X_i is the sum of the variances. If we drop the assumption of independence, we can still say something.

20.4 Theorem

Let $X_1, \ldots, X_n$ be arbitrary random variables (with finite mean and variance). Then

$$\text{Var}(X_1 + \cdots + X_n) = \sum_{i=1}^{n} \text{Var}\, X_i + 2 \sum_{\substack{i,j=1 \\ i<j}}^{n} \text{Cov}(X_i, X_j).$$

For example, the variance of $X_1 + X_2 + X_3 + X_4$ is

$$\sum_{i=1}^{4} \text{Var}\, X_i + 2[\text{Cov}(X_1, X_2) + \text{Cov}(X_1, X_3) + \text{Cov}(X_1, X_4)$$

$$+ \text{Cov}(X_2, X_3) + \text{Cov}(X_2, X_4) + \text{Cov}(X_3, X_4)].$$

Proof. We have

$$E[(X_1 - \mu_1) + \cdots + (X_n - \mu_n)]^2 = E\left[\sum_{i=1}^{n}(X_i - \mu_i)^2\right]$$

$$+2E\left[\sum_{i<j}(X_i - \mu_i)(X_j - \mu_j)\right]$$

as asserted. ♣

The reason for the $i < j$ restriction in the summation can be seen from an expansion such as

$$(x + y + z)^2 = x^2 + y^2 + z^2 + 2xy + 2xz + 2yz.$$

It is correct, although a bit inefficient, to replace $i < j$ by $i \neq j$ and drop the factor of 2. This amounts to writing $2xy$ as $xy + yx$.

20.5 Least Squares

Let $(x_1, y_1), \ldots, (x_n, y_n)$ be points in the plane. The problem is to find the line $y = ax + b$ that minimizes $\sum_{i=1}^{n}[y_i - (ax_i + b)]^2$. (The numbers a and b are to be determined.)

Consider the following random experiment. Choose X with $P\{X = x_i\} = 1/n$ for $i = 1, \ldots, x_n$. If $X = x_i$, set $Y = y_i$. [This is the same setup as in the special case of the Cauchy-Schwarz inequality in (20.3).] Then

$$E[(Y - (aX + b))^2] = \frac{1}{n}\sum_{i=1}^{n}[y_i - (ax_i + b)]^2$$

so the least squares problem is equivalent to finding the best estimate of Y of the form $aX + b$, where "best" means that the mean square error is to be minimized. This is the problem that we solved in (20.1). The least squares line is

$$\boxed{y - \mu_Y = \rho\frac{\sigma_Y}{\sigma_X}(x - \mu_X)}$$

To evaluate $\mu_X, \mu_Y, \sigma_X, \sigma_Y, \rho$:

$$\mu_X = \frac{1}{n}\sum_{i=1}^{n} x_i = \bar{x}, \quad \mu_Y = \frac{1}{n}\sum_{i=1}^{n} y_i = \bar{y},$$

$$\sigma_X^2 = E[(X - \mu_X)^2] = \frac{1}{n}\sum_{i=1}^{n}(x_i - \bar{x})^2 = s_x^2, \quad \sigma_Y^2 = \frac{1}{n}\sum_{i=1}^{n}(y_i - \bar{y})^2 = s_y^2,$$

$$\rho = \frac{E[(X - \mu_X)(Y - \mu_Y)]}{\sigma_X \sigma_Y}, \quad \rho\frac{\sigma_Y}{\sigma_X} = \frac{E[(X - \mu_X)(Y - \mu_Y)]}{\sigma_X^2}.$$

The last entry is the slope of the least squares line, which after cancellation of $1/n$ in numerator and denominator, becomes

$$\frac{\sum_{i=1}^{n}(x_i - \bar{x})(y_i - \bar{y})}{\sum_{i=1}^{n}(x_i - \bar{x})^2}.$$

If $\rho > 0$, then the least squares line has positive slope, and y tends to increase with x. If $\rho < 0$, then the least squares line has negative slope and y tends to decrease as x increases.

Problems

In Problems 1-5, assume that X and Y are independent random variables, and that we know $\mu_X = E(X), \mu_Y = E(Y), \sigma_X^2 = \text{Var } X$, and $\sigma_Y^2 = \text{Var } Y$. In Problem 2, we also know ρ, the correlation coefficient between X and Y.

1. Find the variance of XY.
2. Find the variance of $aX + bY$, where a and b are arbitrary real numbers.
3. FInd the covariance of X and $X + Y$.
4. FInd the correlation coefficient between X and $X + Y$.
5. FInd the covariance of XY and X.
6. Under what conditions will there be equality in the Cauchy-Schwarz inequality?

Lecture 21. The Multivariate Normal Distribution

21.1 Definitions and Comments

The *joint moment-generating function* of $X_1, \ldots, X_n$ [also called the moment-generating function of the random vector $(X_1, \ldots, X_n)$] is defined by

$$M(t_1, \ldots, t_n) = E[\exp(t_1 X_1 + \cdots + t_n X_n)].$$

Just as in the one-dimensional case, the moment-generating function determines the density uniquely. The random variables $X_1, \ldots, X_n$ are said to have the *multivariate normal distribution* or to be *jointly Gaussian* (we also say that the random vector $(X_1, \ldots, X_n)$ is *Gaussian*) if

$$M(t_1, \ldots, t_n) = \exp(t_1 \mu_1 + \cdots + t_n \mu_n) \exp\left(\frac{1}{2} \sum_{i,j=1}^{n} t_i a_{ij} t_j \right)$$

where the t_i and μ_j are arbitrary real numbers, and the matrix A is symmetric and positive definite.

Before we do anything else, let us indicate the notational scheme we will be using. Vectors will be written with an underbar, and are assumed to be column vectors unless otherwise specified. If $\underline{t}$ is a column vector with components $t_1, \ldots, t_n$, then to save space we write $\underline{t} = (t_1, \ldots, t_n)'$. The row vector with these components is the transpose of $\underline{t}$, written $\underline{t}'$. The moment-generating function of jointly Gaussian random variables has the form

$$M(t_1, \ldots, t_n) = \exp(\underline{t}' \underline{\mu}) \exp\left(\frac{1}{2} \underline{t}' A \underline{t} \right).$$

We can describe Gaussian random vectors much more concretely.

21.2 Theorem

Joint Gaussian random variables arise from linear transformations on independent normal random variables.

Proof. Let $X_1, \ldots, X_n$ be independent, with X_i normal $(0, \lambda_i)$, and let $\underline{X} = (X_1, \ldots, X_n)'$. Let $\underline{Y} = B\underline{X} + \underline{\mu}$ where B is nonsingular. Then $\underline{Y}$ is Gaussian, as can be seen by computing the moment-generating function of $\underline{Y}$:

$$M_{\underline{Y}}(\underline{t}) = E[\exp(\underline{t}' \underline{Y})] = E[\exp(\underline{t}' B \underline{X})] \exp(\underline{t}' \underline{\mu}).$$

But

$$E[\exp(\underline{u}' \underline{X})] = \prod_{i=1}^{n} E[\exp(u_i X_i)] = \exp\left(\sum_{i=1}^{n} \lambda_i u_i^2 / 2 \right) = \exp\left(\frac{1}{2} \underline{u}' D \underline{u} \right)$$

where D is a diagonal matrix with λ_i's down the main diagonal. Set $\underline{u} = B'\underline{t}, \underline{u}' = \underline{t}'B$; then

$$M_{\underline{Y}}(t) = \exp(\underline{t}'\underline{\mu})\exp(\frac{1}{2}\underline{t}'BDB'\underline{t})$$

and BDB' is symmetric since D is symmetric. Since $\underline{t}'BDB'\underline{t} = \underline{u}'D\underline{u}$, which is greater than 0 except when $\underline{u} = \underline{0}$ (equivalently when $\underline{t} = \underline{0}$ because B is nonsingular), BDB' is positive definite, and consequently $\underline{Y}$ is Gaussian.

Conversely, suppose that the moment-generating function of $\underline{Y}$ is $\exp(\underline{t}'\underline{\mu})\exp[(1/2)\underline{t}'A\underline{t})]$ where A is symmetric and positive definite. Let L be an orthogonal matrix such that $L'AL = D$, where D is the diagonal matrix of eigenvalues of A. Set $\underline{X} = L'(\underline{Y} - \underline{\mu})$, so that $\underline{Y} = \underline{\mu} + L\underline{X}$. The moment-generating function of $\underline{X}$ is

$$E[\exp(\underline{t}'\underline{X})] = \exp(-\underline{t}'L'\underline{\mu})E[\exp(\underline{t}'L'\underline{Y})].$$

The last term is the moment-generating function of $\underline{Y}$ with $\underline{t}'$ replaced by $\underline{t}'L'$, or equivalently, $\underline{t}$ replaced by $L\underline{t}$. Thus the moment-generating function of $\underline{X}$ becomes

$$\exp(-\underline{t}'L'\underline{\mu})\exp(\underline{t}'L'\underline{\mu})\exp\left(\frac{1}{2}\underline{t}'L'AL\underline{t}\right)$$

This reduces to

$$\exp\left(\frac{1}{2}\underline{t}'D\underline{t}\right) = \exp\left(\frac{1}{2}\sum_{i=1}^{n}\lambda_i t_i^2\right).$$

Therefore the X_i are independent, with X_i normal $(0, \lambda_i)$. ♣

21.3 A Geometric Interpretation

Assume for simplicity that all random variables have zero mean, so that the covariance of U and V is $E(UV)$, which can be regarded as an inner product. Then $Y_1, \ldots, Y_n$ span an n-dimensional space, and $X_1, \ldots, X_n$ is an orthogonal basis for that space. We will see later in the lecture that orthogonality is equivalent to independence. (Orthogonality means that the X_i are uncorrelated, i.e., $E(X_iX_j) = 0$ for $i \neq j$.)

21.4 Theorem

Let $\underline{Y} = \underline{\mu} + L\underline{X}$ as in the proof of (21.2), and let A be the symmetric, positive definite matrix appearing in the moment-generating function of the Gaussian random vector $\underline{Y}$. Then $E(Y_i) = \mu_i$ for all i, and furthermore, A is the *covariance matrix* of the Y_i, in other words, $a_{ij} = \text{Cov}(Y_i, Y_j)$ (and $a_{ii} = \text{Cov}(Y_i, Y_i) = \text{Var } Y_i$).

It follows that the means of the Y_i and their covariance matrix determine the moment-generating function, and therefore the density.

Proof. Since the X_i have zero mean, we have $E(Y_i) = \mu_i$. Let K be the covariance matrix of the Y_i. Then K can be written in the following peculiar way:

$$K = E\left\{\begin{bmatrix} Y_1 - \mu_1 \\ \vdots \\ Y_n - \mu_n \end{bmatrix}(Y_1 - \mu_1, \ldots, Y_n - \mu_n)\right\}.$$

Note that if a matrix M is n by 1 and a matrix N is 1 by n, then MN is n by n. In this case, the ij entry is $E[(Y_i - \mu_i)(Y_j - \mu_j)] = \text{Cov}(Y_i, Y_j)$. Thus

$$K = E[(\underline{Y} - \underline{\mu})(\underline{Y} - \underline{\mu})'] = E(L\underline{X}\underline{X}'L') = LE(\underline{X}\underline{X}')L'$$

since expectation is linear. [For example, $E(M\underline{X}) = ME(\underline{X})$ because $E(\sum_j m_{ij}X_j) = \sum_j m_{ij}E(X_j)$.] But $E(\underline{X}\underline{X}')$ is the covariance matrix of the X_i, which is D. Therefore $K = LDL' = A$ (because $L'AL = D$). ♣

21.5 Finding the Density

From $\underline{Y} = \underline{\mu} + L\underline{X}$ we can calculate the density of $\underline{Y}$. The Jacobian of the transformation from $\underline{X}$ to $\underline{Y}$ is $\det L = \pm 1$, and

$$f_{\underline{X}}(x_1, \ldots, x_n) = \frac{1}{(\sqrt{2\pi})^n} \frac{1}{\sqrt{\lambda_1 \cdots \lambda_n}} \exp\left(-\sum_{i=1}^{n} x_i^2/2\lambda_i\right).$$

We have $\lambda_1 \cdots \lambda_n = \det D = \det K$ because $\det L = \det L' = \pm 1$. Thus

$$f_{\underline{X}}(x_1, \ldots, x_n) = \frac{1}{(\sqrt{2\pi})^n \sqrt{\det K}} \exp\left(-\frac{1}{2}\underline{x}D^{-1}\underline{x}\right).$$

But $\underline{y} = \underline{\mu} + L\underline{x}$, $\underline{x} = L'(\underline{y} - \underline{\mu})$, $\underline{x}'D^{-1}\underline{x} = (\underline{y} - \underline{\mu})'LD^{-1}L'(\underline{y} - \underline{\mu})$, and [see the end of (21.4)] $K = LDL'$, $K^{-1} = LD^{-1}L'$. The density of $\underline{Y}$ is

$$f_{\underline{Y}}(y_1, \ldots, y_n) = \frac{1}{(\sqrt{2\pi})^n \sqrt{\det K}} \exp\left[-\frac{1}{2}(\underline{y} - \underline{\mu})'K^{-1}(\underline{y} - \underline{\mu})\right].$$

21.6 Individually Gaussian Versus Jointly Gaussian

If $X_1, \ldots, X_n$ are jointly Gaussian, then each X_i is normally distributed (see Problem 4), but *not conversely*. For example, let X be normal $(0,1)$ and flip an unbiased coin. If the coin shows heads, set $Y = X$, and if tails, set $Y = -X$. Then Y is also normal $(0,1)$ since

$$P\{Y \leq y\} = \frac{1}{2}P\{X \leq y\} + \frac{1}{2}P\{-X \leq y\} = P\{X \leq y\}$$

because $-X$ is also normal $(0,1)$. Thus $F_X = F_Y$. But with probability $1/2$, $X + Y = 2X$, and with probability $1/2$, $X + Y = 0$. Therefore $P\{X + Y = 0\} = 1/2$. If X and Y were jointly Gaussian, then $X + Y$ would be normal (Problem 4). We conclude that X and Y are individually Gaussian but not jointly Gaussian.

21.7 Theorem

If $X_1, \ldots, X_n$ are jointly Gaussian and uncorrelated $(\text{Cov}(X_i, X_j) = 0$ for all $i \neq j)$, then the X_i are independent.

Proof. The moment-generating function of $\underline{X} = (X_1, \ldots, X_n)$ is

$$M_{\underline{X}}(\underline{t}) = \exp(\underline{t}'\underline{\mu}) \exp\left(\frac{1}{2}\underline{t}'K\underline{t}\right)$$

where K is a diagonal matrix with entries $\sigma_1^2, \sigma_2^2, \ldots, \sigma_n^2$ down the main diagonal, and 0's elsewhere. Thus

$$M_{\underline{X}}(\underline{t}) = \prod_{i=1}^{n} \exp(t_i \mu_i) \exp\left(\frac{1}{2}\sigma_i^2 t_i^2\right)$$

which is the joint moment-generating function of independent random variables $X_1, \ldots, X_n$, whee X_i is normal (μ_i, σ_i^2). ♣

21.8 A Conditional Density

Assume $X_1, \ldots, X_n$ be jointly Gaussian. We find the conditional density of X_n given $X_1, \ldots, X_{n-1}$:

$$f(x_n | x_1, \ldots, x_{n-1}) = \frac{f(x_1, \ldots, x_n)}{f(x_1, \ldots, x_{n-1})}$$

with

$$f(x_1, \ldots, x_n) = (2\pi)^{-n/2} (\det K)^{-1/2} \exp\left[-\frac{1}{2}\sum_{i,j=1}^{n} y_i q_{ij} y_j\right]$$

where $Q = K^{-1} = [q_{ij}], y_i = x_i - \mu_i$. Also,

$$f(x_1, \ldots, x_{n-1}) = \int_{-\infty}^{\infty} f(x_1, \ldots, x_{n-1}, x_n)\, dx_n = B(y_1, \ldots, y_{n-1}).$$

Now

$$\sum_{i,j=1}^{n} y_i q_{ij} y_j = \sum_{i,j=1}^{n-1} y_i q_{ij} y_j + y_n \sum_{j=1}^{n-1} q_{nj} y_j + y_n \sum_{i=1}^{n-1} q_{in} y_i + q_{nn} y_n^2.$$

Thus the conditional density has the form

$$\frac{A(y_1, \ldots, y_{n-1})}{B(y_1, \ldots, y_{n-1})} \exp[-(C y_n^2 + D(y_1, \ldots, y_{n-1}) y_n]$$

with $C = (1/2)q_{nn}$, $D = \sum_{j=1}^{n-1} q_{nj} y_j = \sum_{i=1}^{n-1} q_{in} y_i$ since $Q = K^{-1}$ is symmetric. The conditional density may now be expressed as

$$\frac{A}{B} \exp\left(\frac{D^2}{4C}\right) \exp\left[-C(y_n + \frac{D}{2C})^2\right].$$

We conclude that

> given $X_1, \ldots, X_{n-1}$, X_n is normal.

The conditional variance of X_n (the same as the conditional variance of $Y_n = X_n - \mu_n$) is

$$\frac{1}{2C} = \frac{1}{q_{nn}} \quad \text{because} \quad \frac{1}{2\sigma^2} = C, \sigma^2 = \frac{1}{2C}.$$

Thus

$$\text{Var}(X_n | X_1, \ldots, X_{n-1}) = \frac{1}{q_{nn}}$$

and the conditional mean of Y_n is

$$-\frac{D}{2C} = -\frac{1}{q_{nn}} \sum_{j=1}^{n-1} q_{nj} Y_j$$

so the conditional mean of X_n is

$$E(X_n | X_1, \ldots, X_{n-1}) = \mu_n - \frac{1}{q_{nn}} \sum_{j=1}^{n-1} q_{nj}(X_j - \mu_j).$$

Recall from Lecture 18 that $E(Y|X)$ is the best estimate of Y based on X, in the sense that the mean square error is minimized. In the joint Gaussian case, the best estimate of X_n based on $X_1, \ldots, X_{n-1}$ is linear, and it follows that the best linear estimate is in fact the best overall estimate. This has important practical applications, since linear systems are usually much easier than nonlinear systems to implement and analyze.

Problems

1. Let K be the covariance matrix of *arbitrary* random variables $X_1, \ldots, X_n$. Assume that K is nonsingular to avoid degenerate cases. Show that K is symmetric and positive definite. What can you conclude if K is singular?

2. If $\underline{X}$ is a Gaussian n-vector and $\underline{Y} = A\underline{X}$ with A nonsingular, show that $\underline{Y}$ is Gaussian.

3. If $X_1, \ldots, X_n$ are jointly Gaussian, show that $X_1, \ldots, X_m$ are jointly Gaussian for $m \leq n$.

4. If $X_1, \ldots, X_n$ are jointly Gaussian, show that $c_1 X_1 + \cdots + c_n X_n$ is a normal random variable (assuming it is nondegenerate, i.e., not identically constant).

Lecture 22. The Bivariate Normal Distribution

22.1 Formulas

The general formula for the n-dimensional normal density is

$$f_{\underline{X}}(x_1, \ldots, x_n) = \frac{1}{(\sqrt{2\pi})^n} \frac{1}{\sqrt{\det K}} \exp\left[-\frac{1}{2}(\underline{x} - \underline{\mu})' K^{-1}(\underline{x} - \underline{\mu}) \right]$$

where $E(\underline{X}) = \underline{\mu}$ and K is the covariance matrix of $\underline{X}$. We specialize to the case $n = 2$:

$$K = \begin{bmatrix} \sigma_1^2 & \sigma_{12} \\ \sigma_{12} & \sigma_2^2 \end{bmatrix} = \begin{bmatrix} \sigma_1^2 & \rho\sigma_1\sigma_2 \\ \rho\sigma_1\sigma_2 & \sigma_2^2 \end{bmatrix}, \quad \sigma_{12} = \mathrm{Cov}(X_1, X_2);$$

$$K^{-1} = \frac{1}{\sigma_1^2\sigma_2^2(1 - \rho^2)} \begin{bmatrix} \sigma_2^2 & -\rho\sigma_1\sigma_2 \\ -\rho\sigma_1\sigma_2 & \sigma_1^2 \end{bmatrix} = \frac{1}{1 - \rho^2} \begin{bmatrix} 1/\sigma_1^2 & -\rho/\sigma_1\sigma_2 \\ -\rho/\sigma_1\sigma_2 & 1/\sigma_2^2 \end{bmatrix}.$$

Thus the joint density of X_1 and X_2 is

$$\frac{1}{2\pi\sigma_1\sigma_2\sqrt{1 - \rho^2}} \exp\left\{ -\frac{1}{2(1 - \rho^2)} \left[\left(\frac{x_1 - \mu_1}{\sigma_1}\right)^2 - 2\rho\left(\frac{x_1 - \mu_1}{\sigma_1}\right)\left(\frac{x_2 - \mu_2}{\sigma_2}\right) + \left(\frac{x_2 - \mu_2}{\sigma_2}\right)^2 \right] \right\}$$

The moment-generating function of $\underline{X}$ is

$$M_{\underline{X}}(t_1, t_2) = \exp(\underline{t}'\underline{\mu}) \exp\left(\frac{1}{2}\underline{t}'K\underline{t}\right)$$

$$= \exp\left[t_1\mu_1 + t_2\mu_2 + \frac{1}{2}(\sigma_1^2 t_1^2 + 2\rho\sigma_1\sigma_2 t_1 t_2 + \sigma_2^2 t_2^2) \right].$$

If X_1 and X_2 are jointly Gaussian and uncorrelated, then $\rho = 0$, so that $f(x_1, x_2)$ is the product of a function $g(x_1)$ of x_1 alone and a function $h(x_2)$ of x_2 alone. It follows that X_1 and X_2 are independent. (We proved independence in the general n-dimensional case in Lecture 21.)

From the results at the end of Lecture 21, the conditional distribution of X_2 given X_1 is normal, with

$$E(X_2|X_1 = x_1) = \mu_2 - \frac{q_{21}}{q_{22}}(x_1 - \mu_1)$$

where

$$\frac{q_{21}}{q_{22}} = -\frac{\rho/\sigma_1\sigma_2}{1/\sigma_2^2} = -\frac{\rho\sigma_2}{\sigma_1}.$$

Thus

$$E(X_2|X_1 = x_1) = \mu_2 + \frac{\rho\sigma_2}{\sigma_1}(x_1 - \mu_1)$$

and

$$\mathrm{Var}(X_2|X_1 = x_1) = \frac{1}{q_{22}} = \sigma_2^2(1 - \rho^2).$$

For $E(X_1|X_2 = x_2)$ and $\mathrm{Var}(X_1|X_2 = x_2)$, interchange μ_1 and μ_2, and interchange σ_1 and σ_2.

22.2 Example

Let X be the height of the father, Y the height of the son, in a sample of father-son pairs. Assume X and Y bivariate normal, as found by Karl Pearson around 1900. Assume $E(X) = 68$ (inches), $E(Y) = 69$, $\sigma_X = \sigma_Y = 2, \rho = .5$. (We expect ρ to be positive because on the average, the taller the father, the taller the son.

Given $X = 80$ (6 feet 8 inches), Y is normal with mean

$$\mu_Y + \frac{\rho \sigma_Y}{\sigma_X}(x - \mu_X) = 69 + .5(80 - 68) = 75$$

which is 6 feet 3 inches. The variance of Y given $X = 80$ is

$$\sigma_Y^2(1 - \rho^2) = 4(3/4) = 3.$$

Thus the son will tend to be of above average height, but not as tall as the father. This phenomenon is often called *regression*, and the line $y = \mu_Y + (\rho \sigma_Y / \sigma_X)(x - \mu_X)$ is called the *line of regression* or the *regression line*.

Problems

1. Let X and Y have the bivariate normal distribution. The following facts are known: $\mu_X = -1, \sigma_X = 2$, and the best estimate of Y based on X, i.e., the estimate that minimizes the mean square error, is given by $3X + 7$. The minimum mean square error is 28. Find μ_X, σ_Y and the correlation coefficient ρ between X and Y.

2. Show that the bivariate normal density belongs to the exponential class, and find the corresponding complete sufficient statistic.

Lecture 23. Cramér-Rao Inequality

23.1 A Strange Random Variable

Given a density $f_\theta(x)$, $-\infty < x < \infty$, $a < \theta < b$. We have found maximum likelihood estimates by computing $\frac{\partial}{\partial\theta} \ln f_\theta(x)$. If we replace x by X, we have a random variable. To see what is going on, let's look at a discrete example. If X takes on values x_1, x_2, x_3, x_4 with $p(x_1) = .5, p(x_2) = p(x_3) = .2, p(x_4) = .1$, then $p(X)$ is a random variable with the following distribution:

$$P\{p(X) = .5\} = .5, \quad P\{p(X) = .2\} = .4, \quad P\{p(X) = .1\} = .1$$

For example, if $X = x_2$ then $p(X) = p(x_2) = .2$, and if $X = x_3$ then $p(X) = p(x_3) = .2$. The total probability that $p(X) = .2$ is .4.

The continuous case is, at first sight, easier to handle. If X has density f and $X = x$, then $f(X) = f(x)$. But what is the density of $f(X)$? We will not need the result, but the question is interesting and is considered in Problem 1.

The following two lemmas will be needed to prove the Cramér-Rao inequality, which can be used to compute uniformly minimum variance unbiased estimates. In the calculations to follow, we are going to assume that all differentiations under the integral sign are legal.

23.2 Lemma

$$E_\theta\left[\frac{\partial}{\partial\theta} \ln f_\theta(X)\right] = 0.$$

Proof. The expectation is

$$\int_{-\infty}^{\infty} \left[\frac{\partial}{\partial\theta} \ln f_\theta(x)\right] f_\theta(x)\, dx = \int_{-\infty}^{\infty} \frac{1}{f_\theta(x)} \frac{\partial f_\theta(x)}{\partial\theta} f_\theta(x)\, dx$$

which reduces to

$$\frac{\partial}{\partial\theta} \int_{-\infty}^{\infty} f_\theta(x)\, dx = \frac{\partial}{\partial\theta}(1) = 0. \quad \clubsuit$$

23.3 Lemma

Let $Y = g(X)$ and assume $E_\theta(Y) = k(\theta)$. If $k'(\theta) = dk(\theta)/d\theta$, then

$$k'(\theta) = E_\theta\left[Y \frac{\partial}{\partial\theta} \ln f_\theta(X)\right].$$

Proof. We have

$$k'(\theta) = \frac{\partial}{\partial\theta} E_\theta g(X) = \frac{\partial}{\partial\theta} \int_{-\infty}^{\infty} g(x) f_\theta(x)\, dx = \int_{-\infty}^{\infty} g(x) \frac{\partial f_\theta(x)}{\partial\theta}\, dx$$

93

$$= \int_{-\infty}^{\infty} g(x)\frac{\partial f_\theta(x)}{\partial\theta}\cdot\frac{1}{f_\theta(x)}f_\theta(x)\,dx = \int_{-\infty}^{\infty} g(x)\Big[\frac{\partial}{\partial\theta}\ln f_\theta(x)\Big]f_\theta(x)\,dx$$

$$= E_\theta[g(X)\frac{\partial}{\partial\theta}\ln f_\theta(X)] = E_\theta[Y\frac{\partial}{\partial\theta}\ln f_\theta(X)]. \quad \clubsuit$$

23.4 Cramér-Rao Inequality

Under the assumptions of (23.3), we have

$$\mathrm{Var}_\theta\, Y \geq \frac{[k'(\theta)]^2}{E_\theta\big[\big(\frac{\partial}{\partial\theta}\ln f_\theta(X)\big)^2\big]}.$$

Proof. By the Cauchy-Schwarz inequality,

$$[\mathrm{Cov}(V,W)]^2 = (E[(V-\mu_V)(W-\mu_W)])^2 \leq \mathrm{Var}\,V\,\mathrm{Var}\,W$$

hence

$$[\mathrm{Cov}_\theta(Y,\frac{\partial}{\partial\theta}\ln f_\theta(X))]^2 \leq \mathrm{Var}_\theta\, Y\,\mathrm{Var}_\theta\,\frac{\partial}{\partial\theta}\ln f_\theta(X).$$

Since $E_\theta[(\partial/\partial\theta)\ln f_\theta(X)] = 0$ by (23.2), this becomes

$$(E_\theta[Y\frac{\partial}{\partial\theta}\ln f_\theta(X)])^2 \leq \mathrm{Var}_\theta\, Y\, E_\theta[(\frac{\partial}{\partial\theta}\ln f_\theta(X))^2].$$

By (23.3), the left side is $[k'(\theta)]^2$, and the result follows. $\quad\clubsuit$

23.5 A Special Case

Let $X_1,\dots,X_n$ be iid, each with density $f_\theta(x)$, and take $X = (X_1,\dots,X_n)$. Then $f_\theta(x_1,\dots,x_n) = \prod_{i=1}^n f_\theta(x_i)$ and by (23.2),

$$E_\theta\big[\big(\frac{\partial}{\partial\theta}\ln f_\theta(X)\big)^2\big] = \mathrm{Var}_\theta\,\frac{\partial}{\partial\theta}\ln f_\theta(X) = \mathrm{Var}_\theta\sum_{i=1}^n\frac{\partial}{\partial\theta}\ln f_\theta(X_i)$$

$$= n\,\mathrm{Var}_\theta\,\frac{\partial}{\partial\theta}\ln f_\theta(X_i) = nE_\theta\big[\big(\frac{\partial}{\partial\theta}\ln f_\theta(X_i)\big)^2\big]$$

23.6 Theorem

Let $X_1,\dots,X_n$ be iid, each with density $f_\theta(x)$. If $Y = g(X_1,\dots,X_n)$ is an unbiased estimate of θ, then

$$\mathrm{Var}_\theta\, Y \geq \frac{1}{nE_\theta\big[\big(\frac{\partial}{\partial\theta}\ln f_\theta(X_i)\big)^2\big]}.$$

Proof. Applying (23.5), we have a special case of the Cramér-Rao inequality (23.4) with $k(\theta) = \theta, k'(\theta) = 1.$ ♣

The lower bound in (23.6) is $1/nI(\theta)$, where

$$I(\theta) = E_\theta\Big[\Big(\frac{\partial}{\partial\theta}\ln f_\theta(X_i)\Big)^2\Big]$$

is called the *Fisher information*.

It follows from (23.6) that if Y is an unbiased estimate that meets the Cramér-Rao inequality for all θ (an *efficient estimate*, then Y must be a UMVUE of θ.

23.7 A Computational Simplification

From (23.2) we have

$$\int_{-\infty}^{\infty}\Big(\frac{\partial}{\partial\theta}\ln f_\theta(x)\Big)f_\theta(x)\,dx = 0.$$

Differentiate again to obtain

$$\int_{-\infty}^{\infty}\frac{\partial^2\ln f_\theta(x)}{\partial\theta^2}f_\theta(x)\,dx + \int_{-\infty}^{\infty}\frac{\partial\ln f_\theta(x)}{\partial\theta}\frac{\partial f_\theta(x)}{\partial\theta}\,dx = 0.$$

Thus

$$\int_{-\infty}^{\infty}\frac{\partial^2\ln f_\theta(x)}{\partial\theta^2}f_\theta(x)\,dx + \int_{-\infty}^{\infty}\frac{\partial\ln f_\theta(x)}{\partial\theta}\Big[\frac{\partial f_\theta(x)}{\partial\theta}\frac{1}{f_\theta(x)}\Big]f_\theta(x)\,dx = 0.$$

But the term in brackets on the right is $\partial\ln f_\theta(x)/\partial\theta$, so we have

$$\int_{-\infty}^{\infty}\frac{\partial^2\ln f_\theta(x)}{\partial\theta^2}f_\theta(x)\,dx + \int_{-\infty}^{\infty}\Big(\frac{\partial}{\partial\theta}\ln f_\theta(x)\Big)^2 f_\theta(x)\,dx = 0.$$

Therefore

$$E_\theta\Big[\Big(\frac{\partial}{\partial\theta}\ln f_\theta(X_i)\Big)^2\Big] = -E_\theta\Big[\frac{\partial^2\ln f_\theta(X_i)}{\partial\theta^2}\Big].$$

Problems

1. If X is a random variable with density $f(x)$, explain how to find the distribution of the random variable $f(X)$.

2. Use the Cramér-Rao inequality to show that the sample mean is a UMVUE of the true mean in the Bernoulli, normal (with σ^2 known) and Poisson cases.

Lecture 24. Nonparametric Statistics

We wish to make a statistical inference about a random variable X even though we know nothing at all about its underlying distribution.

24.1 Percentiles

Assume F continuous and strictly increasing. If $0 < p < 1$, then the equation $F(x) = p$ has a unique solution ξ_p, so that $P\{X \leq \xi_p\} = p$. When $p = 1/2$, ξ_p is the median; when $p = .3$, ξ_p is the 30-th percentile, and so on.

Let $X_1, \ldots, X_n$ be iid, each with distribution function F, and let $Y_1, \ldots, Y_n$ be the order statistics. We will consider the problem of estimating ξ_p.

24.2 Point Estimates

On the average, np of the observations will be less than ξ_p. (We have n Bernoulli trials, with probability of success $P\{X_i < \xi_p\} = F(\xi_p) = p$.) It seems reasonable to use Y_k as an estimate of ξ_p, where k is approximately np. We can be a bit more precise. The random variables $F(X_1), \ldots, F(X_n)$ are iid, uniform on $(0,1)$ [see (8.5)]. Thus $F(Y_1), \ldots, F(Y_n)$ are the order statistics from a uniform $(0,1)$ sample. We know from Lecture 6 that the density of $F(Y_k)$ is

$$\frac{n!}{(k-1)!(n-k)!} x^{k-1}(1-x)^{n-k}, \quad 0 < x < 1.$$

Therefore

$$E[F(Y_k)] = \int_0^1 \frac{n!}{(k-1)!(n-k)!} x^k (1-x)^{n-k}\,dx = \frac{n!}{(k-1)!(n-k)!}\beta(k+1, n-k+1).$$

Now $\beta(k+1, n-k+1) = \Gamma(k+1)\Gamma(n-k+1)/\Gamma(n+2) = k!(n-k)!/(n+1)!$, and consequently

$$E[F(Y_k)] = \frac{k}{n+1}, \quad 1 \leq k \leq n.$$

Define $Y_0 = -\infty$ and $Y_{n+1} = \infty$, so that

$$E[F(Y_{k+1}) - F(Y_k)] = \frac{1}{n+1}, \quad 0 \leq k \leq n.$$

(Note that when $k = n$, the expectation is $1 - [n/(n+1)] = 1/(n+1)$, as asserted.)

The key point is that on the average, each $[Y_k, Y_{k+1}]$ produces area $1/(n+1)$ under the density f of the X_i. This is true because

$$\int_{Y_k}^{Y_{k+1}} f(x)\,dx = F(Y_{k+1}) - F(Y_k)$$

and we have just seen that the expectation of this quantity is $1/(n+1)$, $k = 0, 1, \ldots, n$. If we want to accumulate area p, set $k/(n+1) = p$, that is, $k = (n+1)p$.

Conclusion: If $(n + 1)p$ is an integer, estimate ξ_p by $Y_{(n+1)p}$.

If $(n + 1)p$ is not an integer, we can use a weighted average. For example, if $p = .6$ and $n = 13$ then $(n + 1)p = 14 \times .6 = 8.4$. Now if $(n + 1)p$ were 8, we would use Y_8, and if $(n + 1)p$ were 9 we would use Y_9. If $(n + 1)p = 8 + \lambda$, we use $(1 - \lambda)Y_8 + \lambda Y_9$. In the present case, $\lambda = .4$, so we use $.6Y_8 + .4Y_9 = Y_8 + .4(Y_9 - Y_8)$.

24.3 Confidence Intervals

Select order statistics Y_i and Y_j, where i and j are (approximately) symmetrical about $(n + 1)p$. Then $P\{Y_i < \xi_p < Y_j\}$ is the probability that the number of observations less than ξ_p is at least i but less than j, i.e., between i and $j - 1$, inclusive. The probability that exactly k observations will be less than ξ_p is $\binom{n}{k}p^k(1 - p)^{n-k}$, hence

$$P\{Y_i < \xi_p < Y_j\} = \sum_{k=i}^{j-1} \binom{n}{k}p^k(1 - p)^{n-k}.$$

Thus (Y_i, Y_j) is a confidence interval for ξ_p, and we can find the confidence level by evaluating the above sum, possibly with the aid of the normal approximation to the binomial.

24.4 Hypothesis Testing

First let's look at a numerical example. The 30-th percentile $\xi_{.3}$ will be less than 68 precisely when $F(\xi_{.3}) < F(68)$, because F is continuous and strictly increasing. Therefore $\xi_{.3} < 68$ iff $F(68) > .3$. Similarly, $\xi_{.3} > 68$ iff $F(68) < .3$, and $\xi_{.3} = 68$ iff $F(68) = .3$. In general,

$$\xi_{p_0} < \xi \iff F(\xi) > p_0, \quad \xi_{p_0} > \xi \iff F(\xi) < p_0$$

and

$$\xi_{p_0} = \xi \iff F(\xi) = p_0.$$

In our numerical example, if $F(68)$ were actually .4, then on the average, 40 percent of the observations will be 68 or less, as opposed to 30 percent if $F(68) = .3$. Thus a larger than expected number of observations less than or equal to 68 will tend to make us reject the hypothesis that the 30-th percentile is exactly 68. In general, our problem will be

$$H_0 : \xi_{p_0} = \xi \quad (\iff F(\xi) = p_0)$$

$$H_1 : \xi_{p_0} < \xi \quad (\iff F(\xi) > p_0)$$

where p_0 and ξ are specified. If Y is the number of observations less than or equal to ξ, we propose to reject H_0 if $Y \geq c$. (If H_1 is $\xi_{p_0} > \xi$, i.e., $F(\xi) < p_0$, we reject if $Y \leq c$.) Note that Y is the number of nonpositive signs in the sequence $X_1 - \xi, \ldots, X_n - \xi$, and for this reason, the terminology *sign test* is used.

Since we are trying to determine whether $F(\xi)$ is equal to p_0 or greater than p_0, we may regard $\theta = F(\xi)$ as the unknown state of nature. The power function of the test is

$$K(\theta) = P_\theta\{Y \geq c\} = \sum_{k=c}^{n} \binom{n}{k} \theta^k (1 - \theta)^{n-k}$$

and in particular, the significance level (probability of a type 1 error) is $\alpha = K(p_0)$.

The above confidence interval estimates and the sign test are *distribution free*, that is, independent of the underlying distribution function F.

Problems are deferred to Lecture 25.

Lecture 25. The Wilcoxon Test

We will need two formulas:

$$\sum_{k=1}^{n} k^2 = \frac{n(n+1)(2n+1)}{6}, \quad \sum_{k=1}^{n} k^3 = \left[\frac{n(n+1)}{2}\right]^2.$$

For a derivation via the calculus of finite differences, see my on-line text "A Course in Commutative Algebra", Section 5.1.

The hypothesis testing problem addressed by the Wilcoxon test is the same as that considered by the sign test, except that:

(1) We are restricted to testing the *median* $\xi_{.5}$.

(2) We assume that $X_1, \ldots, X_n$ are iid and the underlying density is symmetric about the median (so we are not quite nonparametric). There are many situations where we suspect an underlying normal distribution but are not sure. In such cases, the symmetry assumption may be reasonable.

(3) We use the magnitudes as well as the signs of the deviations $X_i - \xi_{.5}$, so the Wilcoxon test should be more accurate than the sign test.

25.1 How The Test Works

Suppose we are testing $H_0 : \xi_{.5} = m$ vs. $H_1 : \xi_{.5} > m$ based on observations $X_1, \ldots, X_n$. We rank the absolute values $|X_i - m|$ from smallest to largest. For example, let $n = 5$ and $X_1 - m = 2.7, X_2 - m = -1.3, X_3 - m = -0.3, X_4 - m = -3.2, X_5 - m = 2.4$. Then

$$|X_3 - m| < |X_2 - m| < |X_5 - m| < |X_1 - m| < |X_4 - m|.$$

Let R_i be the rank of $|X_i - m|$, so that $R_3 = 1, R_2 = 2, R_5 = 3, R_1 = 4, R_4 = 5$. Let Z_i be the sign of $X_i - m$, so that $Z_i = \pm 1$. Then $Z_3 = -1, Z_2 = -1, Z_5 = 1, Z_1 = 1, Z_4 = -1$. The Wilcoxon statistic is

$$W = \sum_{i=1}^{n} Z_i R_i.$$

In this case, $W = -1 - 2 + 3 + 4 - 5 = -1$. Because the density is symmetric about the median, if R_i is given then Z_i is still equally likely to be ± 1, so $(R_1, \ldots, R_n)$ and $(Z_1, \ldots, Z_n)$ are independent. (Note that if R_j is given, the odds about $Z_i (i \neq j)$ are unaffected since the observations $X_1, \ldots, X_n$ are independent.) Now the R_i are simply a permutation of $(1, 2, \ldots, n)$, so

W is a sum of independent random variables V_i where $V_i = \pm i$ with equal probability.

25.2 Properties Of The Wilcoxon Statistic

Under $H_0, E(V_i) = 0$ and $\mathrm{Var}\, V_i = E(V_i^2) = i^2$, so

$$E(W) = \sum_{i=1}^{n} E(V_i) = 0, \quad \mathrm{Var}\, W = \sum_{i=1}^{n} i^2 = \frac{n(n+1)(2n+1)}{6}.$$

The V_i do not have the same distribution, but the central limit theorem still applies because *Liapounov's condition* is satisfied:

$$\frac{\sum_{i=1}^{n} E[|V_i - \mu_i|^3]}{\left(\sum_{i=1}^{n} \sigma_i^2\right)^{3/2}} \to 0 \quad \text{as} \quad n \to \infty.$$

Now the V_i have mean $\mu_i = 0$, so $|V_i - \mu_i|^3 = |V_i|^3 = i^3$ and $\sigma_i^2 = \mathrm{Var}\, V_i = i^2$. Thus the Liapounov fraction is the sum of the first n cubes divided by the 3/2 power of the sum of the first n squares, which is

$$\frac{n^2(n+1)^2/4}{[n(n+1)(2n+1)/6]^{3/2}}.$$

For large n, the numerator is of the order of n^4 and the denominator is of the order of $(n^3)^{3/2} = n^{9/2}$. Therefore the fraction is of the order of $1/\sqrt{n} \to 0$ as $n \to \infty$. By the central limit theorem, $[W - E(W)]/\sigma(W)$ is approximately normal $(0,1)$ for large n, with $E(W) = 0$ and $\sigma^2(W) = n(n+1)(2n+1)/6$.

If the median is larger than its value m under H_0, we expect W to have a positive bias. Thus we reject H_0 if $W \geq c$. (If H_1 were $\xi_{.5} < m$), we would reject if $W \leq c$.) The value of c is determined by our choice of the significance level α.

Problems

1. Suppose we are using a sign test with $n = 12$ observations to decide between the null hypothesis $H_0 : m = 40$ and the alternative $H_1 : m > 40$, whee m is the median. We use the statistic $Y = $ the number of observations that are less than or equal to 40. We reject H_0 if and only if $Y \leq c$. Find the power function $K(p)$ in terms of c and $p = F(40)$, and the probability α of a type 1 error f $c = 2$.

2. Let m be the median of a random variable with density symmetric about m. Using the Wilcoxon test, we are testing $H_0 : m = 160$ vs. $H_1 : m > 160$ based on $n = 16$ observations, which are as follows: 176.9, 158.3, 152.1, 158.8, 172.4, 169.8, 159.7, 162.7, 156.6, 174.5, 184.4, 165.2, 147.8, 177.8, 160.1, 160.5. Compute the Wilcoxon statistic and determine whether H_0 is rejected at the .05 significance level, i.e., the probability of a type 1 error is .05.

3. When n is small, the distribution of W can be found explicitly. Do it for $n = 1, 2, 3$.

Solutions to Problems

Lecture 1

1. $P\{\max(X,Y,Z) \leq t\} = P\{X \leq t \text{ and } Y \leq t \text{ and } Z \leq t\} = P\{X \leq t\}^3$ by independence. Thus the distribution function of the maximum is $(t^6)^3 = t^{18}$, and the density is $18t^{17}, 0 \leq t \leq 1$.

2. See Figure S1.1. We have

$$P\{Z \leq z\} = \int\int_{y \leq zx} f_{XY}(x,y)\, dx\, dy = \int_{x=0}^{\infty}\int_{y=0}^{zx} e^{-x}e^{-y}\, dy\, dx$$

$$F_Z(z) = \int_0^{\infty} e^{-x}(1-e^{-zx})\, dx = 1 - \frac{1}{1+z}, \quad z \geq 0$$

$$f_Z(z) = \frac{1}{(z+1)^2}, \quad z \geq 0$$

$F_Z(z) = f_Z(z) = 0$ for $z < 0$.

3. $P\{Y = y\} = P\{g(X) = y\} = P\{X \in g^{-1}(y)\}$, which is the number of x_i's that map to y, divided by n. In particular, if g is one-to-one, then $p_Y(g(x_i)) = 1/n$ for $i = 1, \ldots, n$.

4. Since the area under the density function must be 1, we have $ab^3/3 = 1$. Then (see Figure S1.2) $f_Y(y) = f_X(y^{1/3})/|dy/dx|$ with $y = x^3, dy/dx = 3x^2$. In dy/dx we substitute $x = y^{1/3}$ to get

$$f_Y(y) = \frac{f_X(y^{1/3})}{3y^{2/3}} = \frac{3}{b^3}\frac{y^{2/3}}{3y^{2/3}} = \frac{1}{b^3}$$

for $0 < y^{1/3} < b$, i.e., $0 < y < b^3$.

5. Let $Y = \tan X$ where X is uniformly distributed between $-\pi/2$ and $\pi/2$. Then (see Figure S1.3)

$$f_Y(y) = \frac{f_X(\tan^{-1} y)}{|dy/dx|_{x=\tan^{-1} y}} = \frac{1/\pi}{\sec^2 x}$$

with $x = \tan^{-1} y$, i.e., $y = \tan x$. But $\sec^2 x = 1 + \tan^2 x = 1 + y^2$, so $f_Y(y) = 1/[\pi(1+y^2)]$, the Cauchy density.

Lecture 2

1. We have $y_1 = 2x_1, y_2 = x_2 - x_1$, so $x_1 = y_1/2, x_2 = (y_1/2) + y_2$, and

$$\frac{\partial(y_1, y_2)}{\partial(x_1, x_2)} = \begin{vmatrix} 2 & 0 \\ -1 & 1 \end{vmatrix} = 2.$$

103

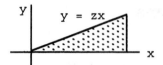

Figure S1.1

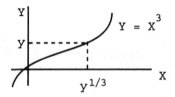

Figure S1.2

Thus $f_{Y_1Y_2}(y_1,y_2) = (1/2)f_{X_1X_2}(x_1,x_2) = e^{-x_1-x_2} = \exp[-(y_1/2)-(y_1/2)-y_2] = e^{-y_1}e^{-y_2}$. As indicated in the comments, the range of the y's is $0 < y_1 < 1, 0 < y_2 < 1$. Therefore the joint density of Y_1 and Y_2 is the product of a function of y_1 alone and a function of y_2 alone, which forces independence.

2. We have $y_1 = x_1/x_2, y_2 = x_2$, so $x_1 = y_1y_2, x_2 = y_2$ and

$$\frac{\partial(x_1,x_2)}{\partial(y_1,y_2)} = \begin{vmatrix} y_2 & y_1 \\ 0 & 1 \end{vmatrix} = y_2.$$

Thus $f_{Y_1Y_2}(y_1,y_2) = f_{X_1X_2}(x_1,x_2)|\partial(x_1,x_2)/\partial(y_1,y_2)| = (8y_1y_2)(y_2)(y_2) = 2y_1(4y_2^3)$. Since $0 < x_1 < x_2 < 1$ is equivalent to $0 < y_1 < 1, 0 < y_2 < 1$, it follows just as in Problem 1 that X_1 and X_2 are independent.

3. The Jacobian $\partial(x_1,x_2,x_3)/\partial(y_1,y_2,y_3)$ is given by

$$\begin{vmatrix} y_2y_3 & y_1y_3 & y_1y_2 \\ -y_2y_3 & y_3-y_1y_3 & y_2-y_1y_2 \\ 0 & -y_3 & 1-y_2 \end{vmatrix}$$

$$= (y_2y_3^2 - y_1y_2y_3^2)(1-y_2) + y_1y_2^2y_3^2 + y_3(y_2-y_1y_2)y_2y_3 + (1-y_2)y_1y_2y_3^2$$

which cancels down to $y_2y_3^2$. Thus

$$f_{Y_1Y_2Y_3}(y_1,y_2,y_3) = \exp[-(x_1+x_2+x_3)]y_2y_3^2 = y_2y_3^2 e^{-y_3}.$$

This can be expressed as $(1)(2y_2)(y_3^2 e^{-y_3}/2)$, and since $x_1, x_2, x_3 > 0$ is equivalent to $0 < y_1 < 1, 0 < y_2 < 1, y_3 > 0$, it follows as before that Y_1, Y_2, Y_3 are independent.

Lecture 3

1. $M_{X_2}(t) = M_Y(t)/M_{X_1}(t) = (1-2t)^{-r/2}/(1-2t)^{-r_1/2} = (1-2t)^{-(r-r_1)/2}$, which is $\chi^2(r-r_1)$.

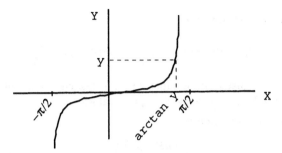

Figure S1.3

2. The moment-generating function of $c_1 X_1 + c_2 X_2$ is

$$E[e^{t(c_1 X_1 + c_2 X_2)}] = E[e^{tc_1 X_1}]E[e^{tc_2 X_2}] = (1 - \beta_1 c_1 t)^{-\alpha_1}(1 - \beta_2 c_2 t)^{-\alpha_2}.$$

If $\beta_1 c_1 = \beta_2 c_2$, then $X_1 + X_2$ is gamma with $\alpha = \alpha_1 + \alpha_2$ and $\beta = \beta_i c_i$.

3. $M(t) = E[\exp(\sum_{i=1}^{n} c_i X_i)] = \prod_{i=1}^{n} E[\exp(tc_i X_i)] = \prod_{i=1}^{n} M_i(c_i t).$

4. Apply Problem 3 with $c_i = 1$ for all i. Thus

$$M_Y(t) = \prod_{i=1}^{n} M_i(t) = \prod_{i=1}^{n} \exp[\lambda_i(e^t - 1)] = \exp\left[\left(\sum_{i=1}^{n} \lambda_i\right)(e^t - 1)\right]$$

which is Poisson $(\lambda_1 + \cdots + \lambda_n)$.

5. Since the coin is unbiased, X_2 has the same distribution as the number of heads in the second experiment. Thus $X_1 + X_2$ has the same distribution as the number of heads in $n_1 + n_2$ tosses, namely binomial with $n = n_1 + n_2$ and $p = 1/2$.

Lecture 4

1. Let Φ be the normal (0,1) distribution function, and recall that $\Phi(-x) = 1 - \Phi(x)$. Then

$$P\{\mu - c < \overline{X} < \mu + c\} = P\{-c\frac{\sqrt{n}}{\sigma} < \frac{\overline{X} - \mu}{\sigma/\sqrt{n}} < c\frac{\sqrt{n}}{\sigma}\}$$

$$= \Phi(c\sqrt{n}/\sigma) - \Phi(-c\sqrt{n}/\sigma) = 2\Phi(c\sqrt{n}/\sigma) - 1 \geq .954.$$

Thus $\Phi(c\sqrt{n}/\sigma) \geq 1.954/2 = .977$. From tables, $c\sqrt{n}/\sigma \geq 2$, so $n \geq 4\sigma^2/c^2$.

2. If $Z = \overline{X} - \overline{Y}$, we want $P\{Z > 0\}$. But Z is normal with mean $\mu = \mu_1 - \mu_2$ and variance $\sigma^2 = (\sigma_1^2/n_1) + (\sigma_2^2/n_2)$. Thus

$$P\{Z > 0\} = P\{\frac{Z - \mu}{\sigma} > \frac{-\mu}{\sigma}\} = 1 - \Phi(-\mu/\sigma) = \Phi(\mu/\sigma).$$

3. Since nS^2/σ^2 is $\chi^2(n-1)$, we have

$$P\{a < S^2 < b\} = P\{\frac{na}{\sigma^2} < \chi^2(n-1) < \frac{nb}{\sigma^2}\}.$$

If F is the $\chi^2(n-1)$ distribution function, the desired probability is $F(nb/\sigma^2) - F(na/\sigma^2)$, which can be found using chi-square tables.

4. The moment-generating function is

$$E[e^{tS^2}] = E\left(\exp\left[\frac{nS^2}{\sigma^2}\frac{t\sigma^2}{n}\right]\right) = E[\exp(t\sigma^2 X/n)]$$

where the random variable X is $\chi^2(n-1)$, and therefore has moment-generating function $M(t) = (1-2t)^{-(n-1)/2}$. Replacing t by $t\sigma^2/n$ we get

$$M_{S^2}(t) = \left(1 - \frac{2t\sigma^2}{n}\right)^{-(n-1)/2}$$

so S^2 is gamma with $\alpha = (n-1)/2$ and $\beta = 2\sigma^2/n$.

Lecture 5

1. By definition of the beta density,

$$E(X) = \frac{\Gamma(a+b)}{\Gamma(a)\Gamma(b)} \int_0^\infty x^a (1-x)^{b-1}\, dx$$

and the integral is $\beta(a+1, b) = \Gamma(a+1)\Gamma(b)/\Gamma(a+b+1)$. Thus $E(X) = a/(a+b)$. Now

$$E(X^2) = \frac{\Gamma(a+b)}{\Gamma(a)\Gamma(b)} \int_0^\infty x^{a+1} (1-x)^{b-1}\, dx$$

and the integral is $\beta(a+2, b) = \Gamma(a+2)\Gamma(b)/\Gamma(a+b+2)$. Thus

$$E(X^2) = \frac{(a+1)a}{(a+b+1)(a+b)}.$$

and

$$\mathrm{Var}\, X = E(X^2) - [E(X)]^2$$

$$= \frac{1}{(a+b)^2(a+b+1)}[(a+1)a(a+b) - a^2(a+b+1)] = \frac{ab}{(a+b)^2(a+b+1)}.$$

2. $P\{-c \le T \le c\} = F_T(c) - F_T(-c) = F_T(c) - (1 - F_T(c)) = 2F_T(c) - 1 = .95$, so $F_T(c) = 1.95/2 = .975$. From the T table, $c = 2.131$.

3. $W = (X_1/m)/(X_2/n)$ where $X_1 = \chi^2(m)$ and $X_2 = \chi^2(n)$. Consequently, $1/W = (X_2/n)/(X_1/m)$, which is $F(n, m)$.

4. Suppose we want $P\{W \le c\} = .05$. Equivalently, $P\{1/W \ge 1/c\} = .05$, hence $P\{1/W \le 1/c\} = .95$. By Problem 3, $1/W$ is $F(n,m)$, so $1/c$ can be found from the F table, and we can then compute c. The analysis is similar for .1, .025 and .01.

5. If N is normal $(0,1)$, then $T(n) = N/(\sqrt{\chi^2(n)/n})$. Thus $T^2(n) = N^2/(\chi^2(n)/n)$. But N^2 is $\chi^2(1)$, and the result follows.

6. If $Y = 2X$ then $f_Y(y) = f_X(x)|dx/dy| = (1/2)e^{-x} = (1/2)e^{-y/2}, y \ge 0$, the chi-square density with two degrees of freedom. If X_1 and X_2 are independent exponential random variables, then

$$\frac{X_1}{X_2} = \frac{(2X_1)/2}{(2X_2)/2} = \frac{\chi^2(2)/2}{\chi^2(2)/2} = F(2,2).$$

Lecture 6

1. Apply the formula for the joint density of Y_j and Y_k with $j = 1, k = 3, n = 3, F(x) = x, f(x) = 1, 0 < x < 1$. The result is $f_{Y_1 Y_3}(x,y) = 6(y - x), 0 < x < y < 1$. Now let $Z = Y_3 - Y_1, W = Y_3$. The Jacobian of the transformation has absolute value 1, so $f_{ZW}(z,w) = f_{Y_1 Y_3}(y_1, y_3) = 6(y_3 - y_1) = 6z, 0 < z < w < 1$. Thus

$$f_Z(z) = \int_{w=z}^{1} 6z \, dw = 6z(1-z), \quad 0 < z < 1.$$

2. The probability that more than one random variable falls in $[x, x + dx]$ need not be negligible. For example, there can be a positive probability that two observations coincide with x.

3. The density of Y_k is

$$f_{Y_k}(x0 = \frac{n!}{(k-1)!(n-k)!}x^{k-1}(1-x)^{n-k}, \quad 0 < x < 1$$

which is beta with $\alpha = k$ and $\beta = n - k + 1$. (Note that $\Gamma(k) = (k-1)!, \Gamma(n-k+1) = (n-k)!, \Gamma(k+n-k+1) = \Gamma(n+1) = n!$.)

4. We have $Y_k > p$ if and only if *at most* $k-1$ observations are in $[0, p]$. But the probability that a particular observation lies in $[0, p]$ is $p/1 = p$. Thus we have n Bernoulli trials with probability of success p on a given trial. Explicitly,

$$P\{Y_k > p\} = \sum_{i=0}^{k-1} \binom{n}{i} p^i (1-p)^{n-i}.$$

Lecture 7

1. Let $W_n = (S_n - E(S_n))/n$; then $E(W_n) = 0$ for all n, and

$$\text{Var } W_n = \frac{\text{Var } S_n}{n^2} = \frac{1}{n^2}\sum_{i=1}^{n}\sigma_i^2 \le \frac{nM}{n^2} = \frac{M}{n} \to 0.$$

It follows that $W_n \xrightarrow{P} 0$.

2. All X_i and X have the same distribution ($p(1) = p(0) = 1/2$), so $X_n \xrightarrow{d} 0$. But if $0 < \epsilon < 1$ then $P\{|X_n - X| \geq \epsilon\} = P\{X_n \neq X\}$, which is 0 for n odd and 1 for n even. Therefore $P\{|X_n - X| \geq \epsilon\}$ oscillates and has no limit as $n \to \infty$.

3. By the weak law of large numbers, $\overline{X}_n$ converges in probability to μ, hence converges in distribution to μ. Thus we can take X to have a distribution function F that is degenerate at μ, in other words,

$$F(x) = \begin{cases} 0, & x < \mu \\ 1, & x \geq \mu. \end{cases}$$

4. Let F_n be the distribution function of X_n. For all x, $F_n(x) = 0$ for sufficiently large n. Since the identically zero function cannot be a distribution function, there is no limiting distribution.

Lecture 8

1. Note that $M_{X_n} = 1/(1-\beta t)^n$ where $1/(1-\beta t)$ is the moment-generating function of an exponential random variable (which has mean β). By the weak law of large numbers, $X_n/n \xrightarrow{P} \beta$, hence $X_n/n \xrightarrow{d} \beta$.

2. $\chi^2(n) = \sum_{i=1}^{n} X_i^2$, where the X_i are iid, each normal $(0,1)$. Thus the central limit theorem applies.

3. We have n Bernoulli trials, with probability of success $p = \int_a^b f(x)\,dx$ on a given trial. Thus Y_n is binomial (n, p). If n and p satisfy the sufficient condition given in the text, the normal approximation with $E(Y_n) = np$ and $\text{Var}\, Y_n = np(1 - p)$ should work well in practice.

4. We have $E(X_i) = 0$ and

$$\text{Var}\, X_i = E(X_i^2) = \int_{-1/2}^{1/2} x^2\, dx = 2 \int_0^{1/2} x^2\, dx = 1/12.$$

By the central limit theorem, Y_n is approximately normal with $E(Y_n) = 0$ and $\text{Var}\, Y_n = n/12$.

5. Let $W_n = n(1 - F(Y_n))$. Then

$$P\{W_n \geq w\} = P\{F(Y_n) \leq 1 - (w/n)\} = P\{\max F(X_i) \leq 1 - (w/n)\}$$

hence

$$P\{W_n \geq w\} = \left(1 - \frac{w}{n}\right)^n, \quad 0 \leq w \leq n,$$

which approaches e^{-w} as $n \to \infty$. Therefore the limiting distribution of W_n is exponential.

Lecture 9

1. (a) We have

$$f_\theta(x_1,\dots,x_n) = \theta^{x_1+\cdots+x_n}\frac{e^{-n\theta}}{x_1!\cdots x_n!}.$$

With $x = x_1 + \cdots + x_n$, take logarithms and differentiate to get

$$\frac{\partial}{\partial\theta}(x\ln\theta - n\theta) = \frac{x}{\theta} - n = 0, \quad \hat\theta = \overline{X}.$$

(b) $f_\theta(x_1,\dots,x_n) = \theta^n(x_1\cdots x_n)^{\theta-1}, \theta > 0$, and

$$\frac{\partial}{\partial\theta}\left(n\ln\theta + (\theta-1)\sum_{i=1}^n \ln x_i\right) = \frac{n}{\theta} + \sum_{i=1}^n \ln x_i = 0, \quad \hat\theta = -\frac{n}{\sum_{i=1}^n}\ln x_i.$$

Note that $0 < x_i < 1$, so $\ln x_i < 0$ for all i and $\hat\theta > 0$.

(c) $f_\theta(x_1,\dots,x_n) = (1/\theta^n)\exp[-(\sum_{i=1}^n x_i)/\theta]$. With $x = \sum_{i=1}^n x_i$ we have

$$\frac{\partial}{\partial\theta}\left(-n\ln\theta - \frac{x}{\theta}\right) = -\frac{n}{\theta} + \frac{x}{\theta^2} = 0, \quad \hat\theta = \overline{X}$$

(d) $f_\theta(x_1,\dots,x_n) = (1/2)^n \exp[-\sum_{i=1}^n |x_i - \theta|]$. We must minimize $\sum_{i=1}^n |x_i - \theta|$, and we must be careful when differentiating because of the absolute values. If the order statistics of the x_i are $y_i, i = 1,\dots,n$, and $y_k < \theta < y_{k+1}$, then the sum to be minimized is

$$(\theta - y_1) + \cdots + (\theta - y_k) + (y_{k+1} - \theta) + \cdots + (y_n - \theta).$$

The derivative of the sum is the number of y_i's less than θ minus the number of y_i's greater than θ. Thus as θ increases, $\sum_{i=1}^n |x_i - \theta|$ decreases until the number of y_i's less than θ equals the number of y_i's greater than θ. We conclude that $\hat\theta$ is the *median* of the X_i.

(e) $f_\theta(x_1,\dots,x_n) = \exp[-\sum_{i=1}^n x_i]e^{n\theta}$ if all $x_i \geq \theta$, and 0 elsewhere. Thus

$$f_\theta(x_1,\dots,x_n) = \exp[-\sum_{i=1}^n x_i]e^{n\theta}I[\theta \leq \min(x_1,\dots,x_n)].$$

The indicator I prevents us from differentiating blindly. As θ increases, so does $e^{n\theta}$, but if $\theta > \min_i x_i$, the indicator drops to 0. Thus $\hat\theta = \min(X_1,\dots,X_n)$.

2. $f_\theta(x_1,\dots,x_n) = 1$ if $\theta - (1/2) \leq x_i \leq \theta + (1/2)$ for all i, and 0 elsewhere. If $Y_1,\dots,Y_n$ are the order statistics of the X_i, then $f_\theta(x_1,\dots,x_n) = I[y_n - (1/2) \leq \theta \leq y_1 + (1/2)]$, where $y_1 = \min x_i$ and $y_n = \max x_i$. Thus any function $h(X_1,\dots,X_n)$ such that

$$Y_n - \frac{1}{2} \leq h(X_1,\dots,X_n) \leq Y_1 + \frac{1}{2}$$

for all $X_1, \ldots, X_n$) is an MLE of θ. Some solutions are $h = Y_1 + (1/2)$, $h = Y_n - (1/2)$, $h = (Y_1 + Y_n)/2$, $h = (2Y_1 + 4Y_n - 1)/6$ and $h = (4Y_1 + 2Y_n + 1)/6$. In all cases, the inequalities reduce to $Y_n - Y_1 \leq 1$, which is true.

3. (a) X_i is Poisson (θ) so $E(X_i) = \theta$. The method of moments sets $\overline{X} = \theta$, so the estimate of θ is $\theta^* = \overline{X}$, which is consistent by the weak law of large numbers.

(b) $E(X_i) = \int_0^1 \theta x^\theta \, d\theta = \theta/(\theta+1) = \overline{X}, \theta = \theta\overline{X} + \overline{X}$, so

$$\theta^* = \frac{\overline{X}}{1 - \overline{X}} \xrightarrow{P} \frac{\theta/(\theta+1)}{1 - [\theta/(\theta+1)]} = \theta$$

hence θ^* is consistent.

(c) $E(X_i) = \theta = \overline{X}$, so $\theta^* = \overline{X}$, consistent by the weak law of large numbers.

(d) By symmetry, $E(X_i) = \theta$ so $\theta^* = \overline{X}$ as in (a) and (c).

(e) $E(X_i) = \int_\theta^\infty x e^{-(x-\theta)} \, dx = $ (with $y = x - \theta$) $\int_0^\infty (y+\theta)e^{-y} \, dy = 1 + \theta = \overline{X}$. Thus $\theta^* = \overline{X} - 1$ which converges in probability to $(1+\theta) - 1 = \theta$, proving consistency.

4. $P\{X \leq r\} = \int_0^r (1/\theta)e^{-x/\theta} \, dx = [-e^{-x/\theta}]_0^r = 1 - e^{-r/\theta}$. The MLE of θ is $\hat{\theta} = \overline{X}$ [see Problem 1(c)], so the MLE of $1 - e^{-r/\theta}$ is $1 - e^{-r/\overline{X}}$.

5. The MLE of θ is X/n, the relative frequency of success. Since

$$P\{a \leq X \leq b\} = \sum_{k=a}^{b} \binom{n}{k} \theta^k (1-\theta)^{n-k},$$

the MLE of $P\{a \leq X \leq b\}$ is found by replacing θ by X/n in the above summation.

Lecture 10

1. Set $2\Phi(b) - 1$ equal to the desired confidence level. This, along with the table of the normal $(0,1)$ distribution function, determines b. The length of the confidence interval is $2b\sigma/\sqrt{n}$.

2. Set $2F_T(b) - 1$ equal to the desired confidence level. This, along with the table of the $T(n-1)$ distribution function, determines b. The length of the confidence interval is $2bS/\sqrt{n-1}$.

3. In order to compute the expected length of the confidence interval, we must compute $E(S)$, and the key observation is

$$S = \frac{\sigma}{\sqrt{n}} \sqrt{\frac{nS^2}{\sigma^2}} = \frac{\sigma}{\sqrt{n}} \sqrt{\chi^2(n-1)}.$$

If $f(x)$ is the chi-square density with $r = n - 1$ degrees of freedom [see (3.8)], then the expected length is

$$\frac{2b}{\sqrt{n-1}} \frac{\sigma}{\sqrt{n}} \int_0^\infty x^{1/2} f(x) \, dx$$

and an appropriate change of variable reduces the integral to a gamma function which can be evaluated explicitly.

4. We have $E(X_i) = \alpha\beta$ and $\text{Var}(X_i) = \alpha\beta^2$. For large n,

$$\frac{\overline{X} - \alpha\beta}{\sqrt{\alpha\beta}/\sqrt{n}} = \frac{\overline{X} - \mu}{\mu/\sqrt{\alpha n}}$$

is approximately normal $(0,1)$ by the central limit theorem. With $c = 1/\sqrt{\alpha n}$ we have

$$P\{-b < \frac{\overline{X} - \mu}{c\mu} < b\} = \Phi(b) - \Phi(-b) = 2\Phi(b) - 1$$

and if we set this equal to the desired level of confidence, then b is determined. The confidence interval is given by $(1 - bc)\mu < \overline{X} < (1 + bc)\mu$, or

$$\frac{\overline{X}}{1 + bc} < \mu < \frac{\overline{X}}{1 - bc}$$

where $c \to 0$ as $n \to \infty$.

5. A confidence interval of length L corresponds to $|(Y_n/n) - p| < L/2$, an event with probability

$$2\Phi\left(\frac{L\sqrt{n}/2}{\sqrt{p(1-p)}}\right) - 1.$$

Setting this probability equal to the desired confidence level gives an inequality of the form

$$\frac{L\sqrt{n}/2}{\sqrt{p(1-p)}} > c.$$

As in the text, we can replace $p(1-p)$ by its maximum value $1/4$. We find the minimum value of n by squaring both sides.

In the first example in (10.1), we have $L = .02$, $L/2 = .01$ and $c = 1.96$. This problem essentially reproduces the analysis in the text in a more abstract form. Specifying how close to p we want our estimate to be (at the desired level of confidence) is equivalent to specifying the length of the confidence interval.

Lecture 11

1. Proceed as in (11.1):

$$Z = \overline{X} - \overline{Y} - (\mu_1 - \mu_2) \quad \text{divided by} \quad \sqrt{\frac{\sigma_1^2}{n} + \frac{\sigma_2^2}{m}}$$

is normal $(0,1)$, and $W = (nS_1^2/\sigma_1^2) + (mS_2^2/\sigma_2^2)$ is $\chi^2(n+m-2)$. Thus $\sqrt{n+m-2}Z/\sqrt{W}$ is $T(n + m - 2)$, but the unknown variances cannot be eliminated.

2. If $\sigma_1^2 = c\sigma_2^2$, then

$$\frac{\sigma_1^2}{n} + \frac{\sigma_2^2}{m} = c\sigma_2^2\left(\frac{1}{n} + \frac{1}{cm}\right)$$

and

$$\frac{nS_1^2}{\sigma_1^2} + \frac{mS_2^2}{\sigma_2^2} = \frac{nS_1^2 + cmS_2^2}{c\sigma_2^2}.$$

Thus σ_2^2 can again be eliminated, and confidence intervals can be constructed, assuming c known.

Lecture 12

1. The given test is an LRT and is completely determined by c, independent of $\theta > \theta_0$.

2. The likelihood ratio is $L(x) = f_1(x)/f_0(x) = (1/4)/(1/6) = 3/2$ for $x = 1, 2$, and $L(x) = (1/8)/(1/6) = 3/4$ for $x = 3, 4, 5, 6$. If $0 \le \lambda < 3/4$, we reject for all x, and $\alpha = 1, \beta = 0$. If $3/4 < \lambda < 3/2$, we reject for $x = 1, 2$ and accept for $x = 3, 4, 5, 6$, with $\alpha = 1/3$ and $\beta = 1/2$. If $3/2 < \lambda \le \infty$, we accept for all x, with $\alpha = 0, \beta = 1$.

   For $\alpha = .1$, set $\lambda = 3/2$, accept when $x = 3, 4, 5, 6$, reject with probability a when $x = 1, 2$. Then $\alpha = (1/3)a = .1, a = .3$ and $\beta = (1/2) + (1/2)(1 - a) = .85$.

3. Since (220-200)/10=2, it follows that when c reaches 2, the null hypothesis is accepted. The associated type 1 error probability is $\alpha = 1 - \Phi(2) = 1 - .977 = .023$. Thus the given result is significant even at the significance level .023. If we were to take additional observations, enough to drive the probability of a type 1 error down to .023, we would still reject H_0. Thus the p-value is a concise way of conveying a lot of information about the test.

Lecture 13

1. We sum $(X_i - np_i)^2/np_i, i = 1, 2, 3$, where the X_i are the observed frequencies and the $np_i = 50, 30, 20$ are the expected frequencies. The chi-square statistic is

$$\frac{(40 - 50)^2}{50} + \frac{(33 - 30)^2}{30} + \frac{(27 - 20)^2}{20} = 2 + .3 + 2.45 = 4.75$$

Since $P\{\chi^2(2) > 5.99\} = .05$ and $4.75 < 5.99$, we accept H_0.

2. The expected frequencies are given by

	A	B	C
1	49	147	98
2	51	153	102

For example, to find the entry in the 2C position, we can multiply the row 2 sum by the column 3 sum and divide by the total number of observations (namely 600) to get

(306)(200)/600=102. Alternatively, we can compute $P(C) = (98 + 102)/600 = 1/3$. We multiply this by the row 2 sum 306 to get $306/3=102$. The chi-square statistic is

$$\frac{(33-49)^2}{49} + \frac{(147-147)^2}{147} + \frac{(114-98)^2}{98} + \frac{(67-51)^2}{51} + \frac{(153-153)^2}{153} + \frac{(86-102)^2}{102}$$

which is $5.224+0+2.612+5.02+0+2.510 = 15.366$. There are $(h-1)(k-1) = 1\times 2 = 2$ degrees of freedom, and $P\{\chi^2(2) > 5.99\} = .05$. Since $15.366 > 5.94$, we reject H_0.

3. The observed frequencies minus the expected frequencies are

$$a - \frac{(a+b)(a+c)}{a+b+c+d} = \frac{ad-bc}{a+b+c+d}, \qquad b - \frac{(a+b)(b+d)}{a+b+c+d} = \frac{bc-ad}{a+b+c+d},$$

$$c - \frac{(a+c)(c+d)}{a+b+c+d} = \frac{bc-ad}{a+b+c+d}, \qquad d - \frac{(c+d)(b+d)}{a+b+c+d} = \frac{ad-bc}{a+b+c+d}.$$

The chi-square statistic is

$$\frac{(ad-bc)^2}{a+b+c+d}\left[\frac{1}{(a+b)(c+d)(a+c)(b+d)}\right] \times$$

$$[(c+d)(b+d) + (a+c)(c+d) + (a+b)(b+d) + (a+b)(a+c)]$$

and the expression in small brackets simplifies to $(a+b+c+d)^2$, and the result follows.

Lecture 14

1. The joint probability function is

$$f_\theta(x_1,\ldots,x_n) = \prod_{i=1}^n \frac{e^{-\theta}\theta^{x_i}}{x_i!} = \frac{e^{-n\theta}\theta^{u(x)}}{x_1!\cdots x_n!}.$$

Take $g(\theta, u(x)) = e^{-n\theta}\theta^{u(x)}$ and $h(x) = 1/(x_1!\cdots x_n!)$.

2. $f_\theta(x_1,\ldots,x_n) = [A(\theta)]^n B(x_1)\cdots B(x_n)$ if $0 < x_i < \theta$ for all i, and 0 elsewhere. This can be written as

$$[A(\theta)]^n \prod_{i=1}^n B(x_i)I\left[\max_{1\le i\le n} x_i < \theta\right]$$

where I is an indicator. We take $g(\theta, u(x)) = A^n(\theta)I[\max x_i < \theta]$ and $h(x) = \prod_{i=1}^n B(x_i)$.

3. $f_\theta(x_1,\ldots,x_n) = \theta^n(1-\theta)^{u(x)}$, and the factorization theorem applies with $h(x) = 1$.

4. $f_\theta(x_1,\ldots,x_n) = \theta^{-n}\exp[-(\sum_{i=1}^n x_i)/\theta]$, and the factorization theorem applies with $h(x) = 1$.

5. $f_\theta(x) = \Gamma(a+b)/[\Gamma(a)\Gamma(b)]x^{a-1}(1-x)^{b-1}$ on $(0,1)$. In this case, $a = \theta$ and $b = 2$. Thus $f_\theta(x) = (\theta+1)\theta x^{\theta-1}(1-x)$, so

$$f_\theta(x_1,\ldots,x_n) = (\theta+1)^n\theta^n \Big(\prod_{i=1}^n x_i\Big)^{\theta-1} \prod_{i=1}^n (1-x_i)$$

and the factorization theorem applies with

$$g(\theta, u(x)) = (\theta+1)^n \theta^n u(x)^{\theta-1}$$

and $h(x) = \prod_{i=1}^n (1-x_i)$.

6. $f_\theta(x) = (1/[\Gamma(\alpha)\beta^\alpha])x^{\alpha-1}e^{-x/\beta}, x > 0$, with $\alpha = \theta$ and β arbitrary. The joint density is

$$f_\theta(x_1,\ldots,x_n) = \frac{1}{[\Gamma(\alpha)]^n \beta^{n\alpha}} u(x)^{\alpha-1} \exp[-\sum_{i=1}^n x_i/\beta]$$

and the factorization theorem applies with $h(x) = \exp[-\sum x_i/\beta]$ and $g(\theta, u(x))$ equal to the remaining factors.

7. We have

$$P_\theta\{X_1' = x_1,\ldots, X_n' = x_n\} = P_\theta\{Y = y\}P\{X_1 = x_1,\ldots, X_n = x_n|Y = y\}$$

We can drop the subscript θ since Y is sufficient, and we can replace X_i' by X_i by definition of B's experiment. The result is

$$P_\theta\{X_1' = x_1,\ldots, X_n' = x_n\} = P_\theta\{X_1 = x_1,\ldots, X_n = x_n\}$$

as desired.

Lecture 17

1. Take $u(X) = X$.

2. The joint density is

$$f_\theta(x_1,\ldots, x_n) = \exp\Big[-\sum_{i=1}^n (x_i - \theta)\Big] I[\min x_i > \theta]$$

so Y_1 is sufficient. Now if $y > \theta$, then

$$P\{Y_1 > y\} = (P\{X_1 > y\})^n = \Big(\int_y^\infty \exp[-(x-\theta)]\,dx\Big)^n = \exp[-n(y-\theta)],$$

so

$$F_{Y_1}(y) = 1 - e^{-n(y-\theta)}, \quad f_{Y_1}(y) = ne^{-n(y-\theta)}, \quad y > \theta.$$

The expectation of $g(Y_1)$ under θ is

$$E_\theta[g(Y_1)] = \int_\theta^\infty g(y)n\exp[-n(y-\theta)]\,dy.$$

If this is 0 for all θ, divide by $e^{n\theta}$ to get

$$\int_\theta^\infty g(y)n\exp(-ny)\,dy = 0.$$

Differentiating with respect to θ, we have $-g(\theta)n\exp(-n\theta) = 0$, so $g(\theta) = 0$ for all θ, proving completeness. The expectation of Y_1 under θ is

$$\int_\theta^\infty yn\exp[-n(y-\theta)]\,dy = \int_\theta^\infty (y-\theta)n\exp[-n(y-\theta)]\,dy + \theta\int_\theta^\infty n\exp[-n(y-\theta)]\,dy$$

$$= \int_0^\infty zn\exp(-nz)\,dz + \theta = \frac{1}{n} + \theta.$$

Thus $E_\theta[Y_1 - (1/n)] = \theta$, so $Y_1 - (1/n)$ is a UMVUE of θ.

3. Since $f_\theta(x) = \theta\exp[(\theta - 1)\ln x]$, the density belongs to the exponential class. Thus $\sum_{i=1}^n \ln X_i$ is a complete sufficient statistic, hence so is $\exp\left[(1/n)\sum_{i=1}^n \ln X_i\right] = u(X_1,\ldots,X_n)$. The key observation is that if Y is sufficient and g is one-to-one, then $g(Y)a$ is also sufficient, since $g(Y)$ conveys exactly the same information as Y does; similarly for completeness.

To compute the maximum likelihood estimate, note that the joint density is $f_\theta(x_1,\ldots,x_n) = \theta^n \exp[(\theta - 1)\sum_{i=1}^n \ln x_i]$. Take logarithms, differentiate with respect to θ, and set the result equal to 0. We get $\hat{\theta} = -n/\sum_{i=1}^n \ln X_i$, which is a function of $u(X_1,\ldots,X_n)$.

4. Each X_i is gamma with $\alpha = 2, \beta = 1/\theta$, so (see Lecture 3) Y is gamma $(2n, 1/\theta)$. Thus

$$E_\theta(1/Y) = \int_0^\infty (1/y)\frac{1}{\Gamma(2n)(1/\theta)^{2n}}y^{2n-1}e^{-\theta y}\,dy$$

which becomes, under the change of variable $z = \theta y$,

$$\frac{\theta^{2n}}{\Gamma(2n)}\int_0^\infty \frac{z^{2n-2}}{\theta^{2n-2}}e^{-z}\frac{dz}{\theta} = \frac{\theta^{2n}}{\theta^{2n-1}}\frac{\Gamma(2n-1)}{\Gamma(2n)} = \frac{\theta}{2n-1}.$$

Therefore $E_\theta[(2n-1)/Y] = \theta$, and $(2n-1)/Y$ is the UMVUE of θ.

5. We have $E(Y_2) = [E(X_1) + E(X_2)]/2 = \theta$, hence $E[E(Y_2|Y_1)] = E(Y_2) = \theta$. By completeness, $E(Y_2|Y_1)$ must be Y_1/n.

6. Since $X_i/\sqrt{\theta}$ is normal $(0,1)$, Y/θ is $\chi^2(n)$, which has mean n and variance $2n$. Thus $E[(Y/\theta)^2] = n^2 + 2n$, so $E(Y^2) = \theta^2(n^2+2n)$. Therefore the UMVUE of θ^2 is $Y^2/(n^2+2n)$.

7. (a) $E[E(I|Y)] = E(I) = P\{X_1 \le 1\}$, and the result follows by completeness.

(b) We compute

$$P\{X_1 = r | X_1 + \cdots + X_n = s\} = \frac{P\{X_1 = r, X_2 + \cdots + X_n = s - r\}}{P\{X_1 + \cdots + X_n\} = s}.$$

The numerator is

$$\frac{e^{-\theta}\theta^r}{r!} e^{-(n-1)\theta} \frac{[(n-1)\theta]^{s-r}}{(s-r)!}$$

and the denominator is

$$\frac{e^{-n\theta}(n\theta)^s}{s!}$$

so the conditional probability is

$$\binom{s}{r} \frac{(n-1)^{s-r}}{n^s} = \binom{s}{r} \left(\frac{n-1}{n}\right)^{s-r} \left(\frac{1}{n}\right)^r$$

which is the probability of r successes in s Bernoulli trials, with probability of success $1/n$ on a given trial. Intuitively, if the sum is s, then each contribution to the sum is equally likely to come from $X_1, \ldots, X_n$.

(c) By (b), $P\{X_1 = 0|Y\} + P\{X_1 = 1|Y\}$ is given by

$$\left(1 - \frac{1}{n}\right)^Y + Y\left(\frac{1}{n}\right)\left(1 - \frac{1}{n}\right)^{Y-1} = \left(\frac{n-1}{n}\right)^Y \left[1 + \frac{Y/n}{(n-1)/n}\right]$$

$$= \left(\frac{n-1}{n}\right)^Y \left[1 + \frac{Y}{n-1}\right].$$

This formula also works for $Y = 0$ because it evaluates to 1.

8. The joint density is

$$f_\theta(x_1, \ldots, x_n) = \frac{1}{\theta_2^n} \exp\left[-\sum_{i=1}^{n} \frac{(x_i - \theta_1)}{\theta_2}\right] I\left[\min_i X_i > \theta_1\right].$$

Since

$$\sum_{i=1}^{n} \frac{(x_i - \theta_1)}{\theta_2} = \frac{1}{\theta_2} \sum_{i=1}^{n} x_i - n\theta_1,$$

the result follows from the factorization theorem.

Lecture 18

1. By (18.4), the numerator of $\delta(x)$ is

$$\int_0^1 \theta \theta^{r-1}(1-\theta)^{s-1}\binom{n}{x}\theta^x(1-\theta)^{n-x}\,d\theta$$

and the denominator is

$$\int_0^1 \theta^{r-1}(1-\theta)^{s-1}\binom{n}{x}\theta^x(1-\theta)^{n-x}\,d\theta.$$

Thus $\delta(x)$ is

$$\frac{\beta(r+x+1, n-x+s)}{\beta(r+x, n-x+s)} = \frac{\Gamma(r+x+1)}{\Gamma(r+x)}\frac{\Gamma(r+s+n)}{\Gamma(r+s+n+1)} = \frac{r+x}{r+s+n}.$$

2. The risk function is

$$E_\theta\left[\left(\frac{r+X}{r+s+n}-\theta\right)^2\right] = \frac{1}{(r+s+n)^2}E_\theta[(X-n\theta+r-r\theta-s\theta)^2]$$

with $E_\theta(X-n\theta)=0$, $E_\theta[(X-n\theta)^2 = \operatorname{Var}X = n\theta(1-\theta)$. Thus

$$R_\delta(\theta) = \frac{1}{(r+s+n)^2}[n\theta(1-\theta) + (r-r\theta-s\theta)^2].$$

The quantity in brackets is

$$n\theta - n\theta^2 + r^2 + r^2\theta^2 + s^2\theta^2 - 2r^2\theta - 2rs\theta + 2rs\theta^2$$

which simplifies to

$$((r+s)^2 - n)\theta^2 + (n - 2r(r+s))\theta + r^2$$

and the result follows.

3. If $r = s = \sqrt{n}/2$, then $(r+s)^2 - n = 0$ and $n - 2r(r+s) = 0$, so

$$R_\delta(\theta) = \frac{r^2}{(r+s+n)^2}.$$

4. The average loss using δ is $B(\delta) = \int_{-\infty}^\infty h(\theta)R_\delta(\theta)\,d\theta$. If $\psi(x)$ has a smaller maximum risk than $\delta(x)$, then since R_δ is constant, we have $R_\psi(\theta) < R_\delta(\theta)$ for all θ. Therefore $B(\psi) < B(\delta)$, contradicting the fact that θ is a Bayes estimate.

Lecture 20

1.

$$\operatorname{Var}(XY) = E[(XY)^2] - (EXEY)^2 = E(X^2)E(Y^2) - (EX)^2(EY)^2$$

$$(\sigma_X^2 + \mu_X^2)(\sigma_Y^2 + \mu_Y^2) - \mu_X^2\mu_Y^2 = \sigma_X^2\sigma_Y^2 + \mu_X^2\sigma_Y^2 + \mu_Y^2\sigma_X^2.$$

2.

$$\text{Var}(aX + bY) = \text{Var}(aX) + \text{Var}(bY) + 2ab\,\text{Cov}(X,Y)$$

$$= a^2\sigma_X^2 + b^2\sigma_Y^2 + 2ab\rho\sigma_X\sigma_Y.$$

3.

$$\text{Cov}(X, X+Y) = \text{Cov}(X,X) + \text{Cov}(X,Y) = \text{Var}\,X + 0 = \sigma_X^2.$$

4. By Problem 3,

$$\rho_{X,X+Y} = \frac{\sigma_X^2}{\sigma_X\sigma_{X+Y}} = \frac{\sigma_X}{\sqrt{\sigma_X^2 + \sigma_Y^2}}.$$

5.

$$\text{Cov}(XY, X) = E(X^2)E(Y) - E(X)^2 E(Y)$$

$$= (\sigma_X^2 + \mu_X^2)\mu_Y - \mu_X^2\mu_Y = \sigma_X^2\mu_Y.$$

6. We can assume without loss of generality that $E(X^2) > 0$ and $E(Y^2) > 0$. We will have equality iff the discriminant $b^2 - 4ac = 0$, which holds iff $h(\lambda) = 0$ for some λ. Equivalently, $\lambda X + Y = 0$ for some λ. We conclude that equality holds if and only if X and Y are *linearly* dependent.

Lecture 21

1. Let $Y_i = X_i - E(X_i)$; then $E[(\sum_{i=1}^n t_i Y_i)^2] \geq 0$ for all $\underline{t}$. But this expectation is

$$E[\sum_i t_i Y_i \sum_j t_j Y_j] = \sum_{i,j} t_i\sigma_{ij}t_j = \underline{t}'K\underline{t}$$

where $\sigma_{ij} = \text{Cov}(X_i, X_j)$. By definition of covariance, K is symmetric, and K is always *nonnegative definite* because $\underline{t}'K\underline{t} \geq 0$ for all $\underline{t}$. Thus all eigenvalues λ_i of K are nonnegative. But $K = LDL'$, so $\det K = \det D = \lambda_1 \cdots \lambda_n$. If K is nonsingular then all $\lambda_i > 0$ and K is positive definite.

2. We have $\underline{X} = C\underline{Z} + \mu$ where C is nonsingular and the Z_i are independent normal random variables with zero mean. Then $\underline{Y} = A\underline{X} = AC\underline{Z} + A\mu$, which is Gaussian.

3. The moment-generating function of $(X_1, \ldots, X_m)$ is the moment-generating function of $(X_1, \ldots, X_n)$ with $t_{m+1} = \cdots = t_n = 0$. We recognize the latter moment-generating function as Gaussian; see (21.1).

4. Let $Y = \sum_{i=1}^n c_i X_i$; then

$$E(e^{tY}) = E\left[\exp\left(\sum_{i=1}^n c_i t X_i\right)\right] = M_{\underline{X}}(c_1 t, \ldots, c_n t)$$

$$= \exp\left(t\sum_{i=1}^{n} c_i\mu_i\right)\exp\left(\frac{1}{2}t^2\sum_{i,j=1}^{n} c_i a_{ij}c_j\right)$$

which is the moment-generating function of a normally distributed random variable. Another method: Let $W = c_1 X_1 + \cdots + c_n X_n = \underline{c}'\underline{X} = \underline{c}'(A\underline{Y} + \underline{\mu})$, where the Y_i are independent normal random variables with zero mean. Thus $W = \underline{b}'\underline{Y} + \underline{c}'\underline{\mu}$ where $\underline{b}' = \underline{c}'A$. But $\underline{b}'\underline{Y}$ is a linear combination of independent normal random variables, hence is normal.

Lecture 22

1. If y is the best estimate of Y given $X = x$, then

$$y - \mu_Y = \frac{\rho\sigma_Y}{\sigma_X}(x - \mu_X)$$

and [see (20.1)] the minimum mean square error is $\sigma_Y^2(1-\rho^2)$, which in this case is 28. We are given that $\rho\sigma_Y/\sigma_X = 3$, so $\rho\sigma_Y = 3 \times 2 = 6$ and $\rho^2 = 36/\sigma_Y^2$. Therefore

$$\sigma_Y^2\left(1 - \frac{36}{\sigma_Y^2}\right) = \sigma_Y^2 - 36 = 28, \quad \sigma_Y = 8, \quad \rho^2 = \frac{36}{64}, \quad \rho = .75.$$

Finally, $y = \mu_Y + 3x - 3\mu_X = \mu_Y + 3x + 3 = 3x + 7$, so $\mu_Y = 4$.

2. The bivariate normal density is of the form

$$f_\theta(x,y) = a(\theta)b(x,y)\exp[p_1(\theta)x^2 + p_2(\theta)y^2 + p_3(\theta)xy + p_4(\theta)x + p_5(\theta)y]$$

so we are in the exponential class. Thus

$$\left(\sum X_i^2, \quad \sum Y_i^2, \quad \sum X_i Y_i, \quad \sum X_i, \quad \sum Y_i\right)$$

is a complete sufficient statistic for $\theta = (\sigma_X^2, \sigma_Y^2, \rho, \mu_X, \mu_Y)$. Note also that any statistic in one-to-one correspondence with this one is also complete and sufficient.

Lecture 23

1. The probability of any event is found by integrating the density on the set defined by the event. Thus

$$P\{a \le f(X) \le b\} = \int_A f(x)\,dx, \quad A = \{x : a \le f(x) \le b\}.$$

2. Bernoulli: $f_\theta(x) = \theta^x(1-\theta)^{1-x}, x = 0, 1$

$$\frac{\partial}{\partial\theta}\ln f_\theta(x) = \frac{\partial}{\partial\theta}[x\ln\theta + (1-x)\ln(1-\theta)] = \frac{x}{\theta} - \frac{1-x}{1-\theta}$$

$$\frac{\partial^2}{\partial\theta^2}\ln f_\theta(x) = -\frac{x}{\theta^2} - \frac{1-x}{(1-\theta)^2}$$

$$I(\theta) = E_\theta\left[\frac{X}{\theta^2} + \frac{1-X}{(1-\theta)^2}\right] = \frac{1}{\theta} + \frac{1}{1-\theta} = \frac{1}{\theta(1-\theta)}$$

since $E_\theta(X) = \theta$. Now

$$\mathrm{Var}_\theta\, Y \geq \frac{1}{nI(\theta)} = \frac{\theta(1-\theta)}{n}.$$

But

$$\mathrm{Var}_\theta\, \overline{X} = \frac{1}{n^2}\,\mathrm{Var}[\mathrm{binomial}(n,\theta)] = \frac{n\theta(1-\theta)}{n^2} = \frac{\theta(1-\theta)}{n}$$

so $\overline{X}$ is a UMVUE of θ.

Normal:

$$f_\theta(x) = \frac{1}{\sqrt{2\pi}\sigma}\exp[-(x-\theta)^2/2\sigma^2]$$

$$\frac{\partial}{\partial\theta}\ln f_\theta(x) = \frac{\partial}{\partial\theta}\left[-\frac{(x-\theta)^2}{2\sigma^2}\right] = \frac{x-\theta}{\sigma^2}$$

$$\frac{\partial^2}{\partial\theta^2}\ln f_\theta(x) = -\frac{1}{\sigma^2}, \quad I(\theta) = \frac{1}{\sigma^2}, \quad \mathrm{Var}_\theta\, Y \geq \frac{\sigma^2}{n}$$

But $\mathrm{Var}_\theta\, \overline{X} = \sigma^2/n$, so $\overline{X}$ is a UMVUE of θ.

Poisson: $f_\theta(x) = e^{-\theta}\theta^x/x!, x = 0, 1, 2\ldots$

$$\frac{\partial}{\partial\theta}\ln f_\theta(x) = \frac{\partial}{\partial\theta}(-\theta + x\ln\theta) = -1 + \frac{x}{\theta}$$

$$\frac{\partial^2}{\partial\theta^2}\ln f_\theta(x) = -\frac{x}{\theta^2}, \quad I(\theta) = E\left(\frac{X}{\theta^2}\right) = \frac{\theta}{\theta^2} = \frac{1}{\theta}$$

$$\mathrm{Var}_\theta\, Y \geq \frac{\theta}{n} = \mathrm{Var}_\theta\, \overline{X}$$

so $\overline{X}$ is a UMVUE of θ.

Lecture 25

1.

$$K(p) = \sum_{k=0}^{c} \binom{n}{k} p^k (1-p)^{n-k}$$

with $c = 2$ and $p = 1/2$ under H_0. Therefore

$$\alpha = \left[\binom{12}{0} + \binom{12}{1} + \binom{12}{2} \right] (1/2)^n = \frac{79}{4096} = .019.$$

2. The deviations, with ranked absolute values in parentheses, are

   16.9(14), -1.7(5), -7.9(9), -1.2(4), 12.4(12), 9.8(10), -.3(2), 2.7(6), -3.4(7), 14.5(13), 24.4(16), 5.2(8), -12.2(11), 17.8(15), .1(1), .5(3)

   The Wilcoxon statistic is $W = 1-2+3-4-5+6-7+8-9+10-11+12+13+14+15+16 = 60$

   Under H_0, $E(W) = 0$ and $\text{Var } W = n(n+1)(2n+1)/6 = 1496$, $\quad \sigma_W = 38.678$

   Now $W/38.678$ is approximately normal $(0,1)$ and $P\{W \geq c\} = P\{W/38.678 \geq c/38.678\} = .05$. From a normal table, $c/38.678 = 1.645$, $c = 63.626$. Since $60 < 63.626$, we accept H_0.

3. The moment-generating function of V_j is $M_{V_j}(t) = (1/2)(e^{-jt} + e^{jt})$ and the moment-generating function of W is $M_W(t) = \prod_{j=1}^{n} M_{V_j}(t)$. When $n = 1$, $W = \pm 1$ with equal probability. When $n = 2$,

$$M_W(t) = \frac{1}{2}(e^{-t} + e^t)\frac{1}{2}(e^{-2t} + e^{2t}) = \frac{1}{4}(e^{-3t} + e^{-t} + e^t + e^{3t})$$

so W takes on the values $-3, -1, 1, 3$ with equal probability. When $n = 3$,

$$M_W(t) = \frac{1}{4}(e^{-3t} + e^{-t} + e^t + e^{3t})\frac{1}{2}(e^{-3t} + e^{3t})$$

$$= \frac{1}{8}(e^{-6t} + e^{-4t} + e^{-2t} + 1 + 1 + e^{2t} + e^{4t} + e^{6t}).$$

Therefoe $P\{W = k\} = 1/8$ for $k = -6, -4, -2, 2, 4, 6$, $P\{W = 0\} = 1/4$, and $P\{W = k\} = 0$ for other values of k

Index